RETHINKING THE DRONE WAR

RETHINKING THE DRONE WAR

NATIONAL SECURITY, LEGITIMACY, AND CIVILIAN CASUALTIES

IN U.S. COUNTERTERRORISM OPERATIONS

LARRY LEWIS

DIANE M. VAVRICHEK

A joint publication of CNA
and Marine Corps University Press

MCUP

Quantico, Virginia

2016

This book represents the best opinion of the authors at the time of printing. The views and opinions are the authors' own, and do not necessarily represent those of CNA, the Department of the Navy, the U.S. Marine Corps, Marine Corps University, or the U.S. government.

MCUP

Marine Corps University Press
111 South Street
Quantico, VA 22134
www.usmcu.edu/mcupress
1st printing, 2016

CNA
ANALYSIS & SOLUTIONS

CNA Corporation
3003 Washington Boulevard
Arlington, VA 22201
www.cna.org

Library of Congress Cataloging-in-Publication Data

Names: Lewis, Larry L. (Lawrence L.), author. | Vavrichek, Diane, author.
Title: Rethinking the drone war : national security, legitimacy, and civilian casualties in U.S. counterterrorism operations / Larry Lewis, Diane Vavrichek.
Other titles: National security, legitimacy, and civilian casualties in U.S. counterterrorism operations
Description: Quantico, VA : CNA and Marine Corps University Press, [2016] | Includes index.
Identifiers: LCCN 2016034013
Subjects: LCSH: Drone aircraft--United States--History. | Drone aircraft--Government policy--United States. | Drone aircraft--Moral and ethical aspects--United States. | Uninhabited combat aerial vehicles--Government policy--United States. | Terrorism--Prevention--Government policy--United States. | Targeted killing--Government policy--United States. | Civilian war casualties--Prevention. | United States--Military policy--Moral and ethical aspects. | United States--Military policy--Public opinion. | National security--United States.
Classification: LCC UG1242.D7 L48 2016 | DDC 363.325/16--dc23
LC record available at https://lccn.loc.gov/2016034013

ISBN 978-0-9973174-3-5

Drone Strikes in Pakistan © 2014 by CNA Corporation

The Future of Drone Strikes © 2014 by CNA Corporation

Improving Lethal Action © 2014 by CNA Corporation

Security and Legitimacy © by CNA Corporation (forthcoming)

CONTENTS

Illustrations — vii

Tables — ix

Preface — xi

Acknowledgments — xiii

Abbreviations — xv

PART I
Drone Strikes in Pakistan: Assessing Civilian Casualties

    Chapter 1: Introduction — 1

    Chapter 2: Drone Strike Casualty Estimates — 6

    Chapter 3: Discrepancies in Civilian Casualty Estimates — 12

    Chapter 4: Platform Precision or Comprehensive Process? — 20

    Chapter 5: The Drone Campaign and Civilian Harm — 24

    Chapter 6: Conclusions and Recommendations — 30

PART II
The Future of Drone Strikes: A Framework for Analyzing Policy Options

    Chapter 7: Introduction — 37

    Chapter 8: Framework and Policy Options — 44

    Chapter 9: Military Effectiveness — 57

    Chapter 10: Legitimacy — 65

| Chapter 11: | Anticipating Net Effectiveness | 89 |
| Chapter 12: | Conclusions and Recommendations | 95 |

PART III
Improving Lethal Action: Learning and Adapting in U.S. Counterterrorism Operations

Chapter 13:	Introduction	105
Chapter 14:	An Analytical Approach to Improve Lethal Action Operations	114
Chapter 15:	Illustrating the Approach	127
Chapter 16:	Implementing the Approach	150
Chapter 17:	Benefits of the Process	155
Chapter 18:	Conclusions and Recommendations	164

PART IV
Security and Legitimacy: Learning from the Past Decade of Operations

Chapter 19:	Introduction	171
Chapter 20:	Lesson One: Promoting Legitimacy	175
Chapter 21:	Lesson Two: Practicing Legitimacy	189
Chapter 22:	Conclusions and Recommendations	197

Appendix A: Title 10, Title 50, and Oversight — 209

Appendix B: Covert Actions — 213

Index — 223

ILLUSTRATIONS

Figure 1.   Disparate estimates for civilian deaths from drone strikes in Pakistan, 2004–13

Figure 2.   Comprehensive process for reducing and mitigating civilian harm

Figure 3.   Number of drone strikes in Pakistan per year

Figure 4.   Percent of operations resulting in civilian deaths per year

Figure 5.   Analytical approach to improve lethal force operations

Figure 6.   Perceived tension between mission success and CIVCAS

Figure 7.   Common relationship between mission success and CIVCAS in Afghanistan

Figure 8.   Relationship between mission success and CIVCAS

Figure 9.   Lethal action operations per year in Pakistan and Yemen

Figure 10. Target type distribution for lethal action operations in Pakistan and Yemen, 2012–14

Figure 11. Percent of lethal action operations that successfully targeted a senior leader, by target type

Figure 12.   Maximum/minimum combatant casualties from lethal action operations with different target types

Figure 13.   Percent of lethal action operations causing civilian deaths, by year

Figure 14.   Average civilian deaths per incident, by year

Figure 15.   Average civilian deaths per operation

Figure 16.   Relationship between mission success and CIVCAS

Figure 17.   Initial locations of U.S. forces and civilian convoy

Figure 18.   Movement of civilian convoy

Figure 19.   Strike on civilian convoy

Appendix figure 1. Relationship between legal and doctrinal definitions of covert

TABLES

Table 1. Overall statistics for drone strikes in Pakistan, 2004–13

Table 2. Potential advantages of increased TME and perceived legitimacy

Table 3. Potential effects of policy options on military effectiveness and identified legitimacy issues

Table 4. Short- and long-term effects of mission success and CIVCAS

Table 5. Hypothetical report card for meeting PPG goals

Table 6. Overall civilian death statistics for Pakistan and Yemen through August 2014

Table 7. Comparing CIVCAS from drone strikes and other operations in Yemen

Table 8. Operations in Yemen and Pakistan achieve the aims of the PPG

Appendix table 1. Definitions of covert

PREFACE

The years since 9/11 have seen an evolution in U.S. national security practices with respect to counterterrorism operations, particularly when it comes to lethal force via the use of armed drones, also known as unmanned aerial vehicles, or UAVs.

Drones constitute a recent addition to the long list of technological advancements in warfighting over time, a list that includes the development of the crossbow, gunpowder-powered projectile weapons, chemical weapons in World War I, and rockets and jet aircraft in World War II. The most advanced drones—the armed Predator and Reaper models—offer persistent surveillance as well as the ability to engage targets from almost anywhere across the globe, without a requirement for forces on the ground. This extended reach offers a clear example of how advances in technology can provide a new and effective option for using lethal force.

Technological advancements and changing security practices affecting the use of force raise important ethical and practical questions, such as whether the technologies and practices should be used in warfare and whether self-imposed limits are appropriate for their responsible use, which can be addressed through national policy decisions, as well as such international laws and standards as the principles of humanity enshrined in the Geneva Conventions of 1949. The modern counterterrorism context is no exception. U.S. national security practices, especially those involving armed drones, have raised numerous questions related to ethics and practicality. This volume is a collection of four reports that collectively address these issues by exploring the themes of legitimacy, civilian protection, and national security interests. They address, for example, the following questions in the modern counterterrorism context.

Legitimacy
Do U.S. means and methods of war enhance our perceived legitimacy and influence? Would we want other nations to model our own behavior?

Civilian Protection
A fundamental tenet of international law governing armed conflict is to safeguard civilians from the effects of war when possible. How effective are we in protecting civilians while being effective against the threat?

National Security Interests
A sovereign nation has both the right and obligation to protect its own citizens and look after their welfare. How do our practices and adoption of lethal force technologies affect these interests in the short and long term?

This volume provides concrete recommendations for policy makers as well as military commanders, a number of which have, since the time of its writing, been incorporated into the recent U.S. policy guidelines related to civilian casualties.

Although this volume focuses specifically on current counterterrorism practices, its analysis, frameworks, and conclusions can be applied in varying degrees to emerging technologies as well. These considerations can help a government ensure that its use of force is not only effective, but also responsible and consonant with its larger interests.

ACKNOWLEDGMENTS

Portions of this work were funded by CNA. CNA, a nonprofit research and analysis organization with more than 600 analysts and professional staff, is dedicated to developing actionable solutions to complex problems of national importance. In addition to defense-related matters for the U.S. Navy and Marine Corps, CNA's research portfolio includes policing, homeland security, climate change, water resources, education, and air traffic management.

The authors gratefully acknowledge the individuals who aided their research by granting interviews; as well as the following colleagues for their feedback and support: William D. Brobst, Richard Brody, Alan C. Brown, Marc Garlasco, Matthew Grund, Lieutenant Commander Matthew Ivey (U.S. Navy, Judge Advocate General's Corps [JAGC]), Captain Mark E. Rosen (U.S. Navy, JAGC, Ret), and Jonathan Schroden.

ABBREVIATIONS

AAR	After action report
AFRICOM	U.S. Africa Command
AQAP	al-Qaeda in the Arabian Peninsula
AQIM	al-Qaeda in the Islamic Maghreb
ASM	Air-to-surface missile
AUMF	Authorization for the Use of Military Force
BDA	Battle damage assessment
BIJ	Bureau of Investigative Journalism
BPC	Building partner capacity
CALL	Center for Army Lessons Learned
CENTCOM	U.S. Central Command
CIA	Central Intelligence Agency
CIVCAS	Civilian casualties
CIV K	Civilians killed
CNA	Center for Naval Analyses

COIN	Counterinsurgency
COMISAF	Commander, International Security Assistance Force
CPA	Coalition Provisional Authority
CT	Counterterrorism
DNI	Director of National Intelligence
DOD	Department of Defense
DOJ	Department of Justice
DPH	Directly participate in hostilities
EOF	Escalation of force
FARC	*Fuerzas Armadas Revolucionarias de Colombia* or the Revolutionary Armed Forces of Colombia—People's Army
FATA	Federally Administered Tribal Areas
FBI	Federal Bureau of Investigation
FISA	Foreign Intelligence Surveillance Act
FMS	Foreign Military Sales
FOIA	Freedom of Information Act
HUMINT	Human intelligence
IAC	International armed conflict
ICRC	International Committee of the Red Cross
IED	Improvised explosive device
IHL	International humanitarian law

IHRL	International human rights law
ISAF	International Security Assistance Force
ISIL	Islamic State of Iraq and the Levant (a.k.a. IS, ISIS)
JCCS	Joint Civilian Casualty Study
JCOA	Joint and Coalition Operational Analysis Division/Joint Center for Operational Analysis
JLLIS	Joint Lessons Learned Information System
KIA	Killed in action
LOAC	Law of armed conflict
MAM	Military-age male
MEK	Mujahedin-e Khalq
MILGROUP	U.S. military group
MNF-I	Multinational Force–Iraq
NAF	New America Foundation
NATO	North Atlantic Treaty Organization
NCT	Nation containing the target
NGO	Nongovernmental organization
NIAC	Noninternational Armed Conflict
NSA	National Security Agency
OEF	Operation Enduring Freedom
OGA	Other (i.e., nonmilitary) government agency

OGC	Office of General Counsel (DOD)
OME	Operational military effectiveness
OPE	Operational preparation of the environment
OSD	Office of the Secretary of Defense
PED	Processing, exploitation, and dissemination
PID	Positive identification
POW	Prisoner of war
PPG	Presidential Policy Guidance
QRF	Quick reaction force
ROE	Rules of engagement
SA	Situational awareness
SERE	Survival, Evasion, Resistance, and Escape training
SME	Strategic military effectiveness
SOF	Special Operations Forces
TLAM	Tomahawk land attack missile
TMA	Traditional military activity
TME	Tactical military effectiveness
TTP	Tactics, techniques, and procedures
UAS	Unmanned aerial system
UAV	Unmanned aerial vehicle ("drone")
UCMJ	Uniformed Code of Military Justice

UN	United Nations
UNSCR	United Nations Security Council Resolution
USAID	U.S. Agency for International Development

PART I

DRONE STRIKES IN PAKISTAN

ASSESSING CIVILIAN CASUALTIES
BY LARRY LEWIS

CHAPTER 1

INTRODUCTION

Drones (referred to as unmanned aerial vehicles or UAVs by the U.S. military) are a recent innovation in warfare, introducing new and important capabilities to the battlefield.[1] For example, drone platforms offer persistence beyond the endurance of manned aircraft, allowing intelligence collection, pattern of life development, and attacks on targets in remote areas using armed drones. Drones also provide intelligence feeds that are distributed to and interpreted by a team of analysts and operators, often at great distances from the platform. In addition, drones offer an integrated option for collecting intelligence and striking targets in other countries without requiring boots on the ground, avoiding force protection concerns as well as more overt infringements of national sovereignty.

Some concerns have been expressed, however, over the increasing use of drones. These concerns include:

---

[1] While this technology is still very young, the origins of UAVs can be seen in the early recoverable and reusable radio-controlled aircraft from the 1930s. For more information, see Chris Cole, "Rise of the Reapers: A Brief History of Drones," Drone Wars UK, 10 June 2014, http://dronewars.net/2014/10/06/rise-of-the-reapers-a-brief-history-of-drones/. The Central Intelligence Agency (CIA) first began working with drone technology in 2000 and armed them following the terrorist attacks of 11 September 2001.

- ethical considerations, such as the effect of the increased distance between the target and the person pulling the trigger on decisions regarding the use of force;[2]

- legal considerations, such as the legal basis for the use of force in areas outside of declared areas of armed conflict;[3]

- civilian casualties and the tactics perceived to be associated with drone use, such as double-tapping targets or signature strikes;[4] and

- the various organizations employing drones and consequent implications for the use of force, such as respective practices for considering collateral damage and whether they comply with international humanitarian law, such as the Geneva Conventions and their additional protocols.[5]

In light of these advantages and concerns, a broad public debate has begun concerning the use of drones. Perhaps the most contentious aspect of that debate involves their use in the U.S. counterterrorism (CT) campaign, and

---

[2] Peter W. Singer, *Can Drones and Viruses Be Ethical Weapons?* (Washington, DC: Brookings, 2014), http://www.brookings.edu/research/interviews/2014/02/28-can-drones-viruses-be-ethical-weapons-singer.

[3] Ben Emmerson, *Promotion and Protection of Human Rights and Fundamental Freedoms while Countering Terrorism*, UN Doc. A/68/389 (New York: UN General Assembly, 2013).

[4] For the purposes of this discussion, double tapping refers to instances when a targeted site is struck multiple times over a short period. Unfortunately, the second strike has a higher probability of targeting first responders or U.S. forces on the ground. For more, see Michael B. Kelley, "More Evidence that Drones Are Targeting Civilian Rescuers in Afghanistan," *Business Insider*, 25 September 2012, http://www.businessinsider.com/drone-double-tap-first-responders-2012-9. For more on signature strikes, see *The Civilian Impact of Drones: Unexamined Costs, Unanswered Questions* (Washington, DC: Human Rights Clinic at Columbia Law School and the Center for Civilians in Conflict, 2012), http://civiliansinconflict.org/resources/pub/the-civilian-impact-of-drones.

[5] Charli Carpenter, "Are CIA Drone Pilots Likelier to Comply with International Law?," *Duck of Minerva* (blog), 22 January 2014, http://duckofminerva.com/2014/01/are-cia-drone-pilots-likelier-to-comply-with-international-law.html.

specifically the "users" ability to discriminate between terrorists and civilians.[6] Part I of this monograph examines that issue.

DRONES IN COUNTERTERRORISM OPERATIONS

The United States uses drone strikes to target members of al-Qaeda, the Taliban, and affiliated terrorist groups. These drone operations have been conducted both in major theaters of operation (e.g., Iraq and Afghanistan) and in CT campaigns outside of declared theaters of operation (e.g., Pakistan and Yemen). The U.S. government justifies these campaigns based on an imminent threat to U.S. interests and the minimal cost of this approach to civilian lives.[7] In sum, U.S. officials describe drone strikes as both effective and "surgical." While there is some controversy regarding the legality of the use of force outside of declared theaters of conflict, the topic of legal use of force will not be explored in this work.

In the short term, drone strikes appear to have weakened enemy networks operating in such areas as Pakistan. In the long term, however, these benefits can be undermined by the tendency of drone strikes to create grievances that can both radicalize populations and increase support for terrorist elements.[8] In addition, civilian casualties from CT operations negatively impact the nations in which operations occur, reinforcing concerns over U.S. encroachment of national sovereignty and creating political pressure for those governments. In response, these nations can limit or discourage the conduct of such operations, hindering the American government's ability to respond to imminent threats over the long term. Thus, civilian

---

[6] Larisa Epatko, "Controversy Surrounds Increased Use of U.S. Drone Strikes," PBS News, 10 October 2011, http://www.pbs.org/newshour/rundown/drone-strikes-1/.

[7] Imminent does not necessarily mean immediate in this context. Under the U.S. interpretation of anticipatory self-defense, the "principle of imminence does not involve a requirement to have clear evidence that a specific attack will be carried out in the immediate future." See Emmerson, *Promotion and Protection of Human Rights.*

[8] For example, a number of British members of Parliament wrote a letter to the U.S. government expressing concerns about the radicalization caused by civilian casualties during drone strikes. "Drone Attacks Lead to Terrorism," Letters to the Editor, *Times* (London), 26 July 2012.

casualties from U.S. operations can simultaneously increase the threat to the nation and reduce our ability to confront them.

While the United States should carefully monitor these concerns, imperatives for immediate action can sometimes trump such long-term considerations. For example, the 2011 raid on Osama bin Laden's Abbottabad compound clearly has had lasting effects on the U.S. relationship with Pakistan, but the value of direct action was regarded as paramount in that particular case.[9]

U.S. officials and some academics have described the precision and low collateral damage nature of drone strikes using such adjectives as "surgical" and "humane."[10] U.S. officials have regularly stated that reducing the risk of civilian casualties is a national priority:

- President Barack H. Obama: "Before any strike is taken, there must be near-certainty that no civilians will be killed or injured—the highest standard we can set."[11]

- Former Deputy National Security Advisor and current CIA Director John O. Brennan: "We've done everything possible in Afghanistan and other areas to reduce any risk to that civilian population."[12]

---

[9] While not accomplished with a drone strike, this raid represents one end of the spectrum with regard to trading off benefits of CT actions with their potential negative second-order effects.

[10] Such descriptions from academia include Daniel L. Byman, *Why Drones Work: The Case for Washington's Weapon of Choice* (Washington, DC: Brookings, 2013), http://www.brookings.edu/research/articles/2013/06/17-drones-obama-weapon-choice-us-counterterrorism-byman; and Michael W. Lewis, "Drones: Actually the Most Humane Form of Warfare Ever," *Atlantic*, 21 August 2013, http://www.theatlantic.com/international/archive/2013/08/drones-actually-the-most-humane-form-of-warfare-ever/278746/.

[11] Barack H. Obama, "Remarks by the President at the National Defense University" (speech, Fort McNair, Washington, DC, 23 May 2013), https://www.whitehouse.gov/the-press-office/2013/05/23/remarks-president-national-defense-university.

[12] Ken Dilanian, "Brennan Defends U.S. Drone Attacks Despite Risks to Civilians," *Los Angeles Times*, 29 April 2012, http://latimesblogs.latimes.com/world_now/2012/04/brennan-drone-attacks.html.

While the U.S. government's stated commitment to minimizing civilian harm is laudable, reported levels of civilian casualties for operations in Pakistan differ significantly from nearly every other estimate available, including several open source estimates and a recent UN report.[13] The disparity between the two sets of civilian casualty estimates—those from the U.S. government versus those from nongovernmental and international organizations—raises two salient questions: who is right, and why is there such a disparity?

Part I of this work explores this disparity and its possible causes, and examines the underlying assertion that drones are inherently surgical in their abilities, creating minimal civilian risk. Open source data will then show trends in civilian casualties from U.S. drone strikes in Pakistan and illustrate ways to identify root causes of these incidents to inform improvement in future operations.

---

[13] Emmerson, *Promotion and Protection of Human Rights*.

CHAPTER 2

DRONE STRIKE CASUALTY ESTIMATES: ARE U.S. GOVERNMENT NUMBERS ACCURATE?

When examining civilian casualties, it is critical to define what a civilian is.[1] For the purpose of this discussion, civilians are "those persons who are not combatants (members of military/paramilitary forces) or members of organized armed groups of a party of a conflict."[2] The term *civilian casualty* refers to the death or injury of a civilian as a result of actions of a combatant entity (e.g., the United States, a Coalition partner, host-nation security forces, or insurgents/terrorists). Note that this represents a negative definition; per international humanitarian law (IHL), the burden of proof is to determine whether a casualty is a combatant. If their identity is in doubt, they should be considered civilian. The consequences of this principle in counting civilians will be discussed later. Also, a civilian casualty incident is defined as an operation where civilian harm is caused. For our purposes, the term *civilian deaths* is also used, because it can be difficult to reliably de-

---

[1] This paper uses language consistent with the U.S. perspective that the CT drone campaign is an armed conflict, so that the legal conventions and operating definitions of an armed conflict apply.

[2] *Afghanistan: Annual Report 2011; Protection of Civilians in Armed Conflict* (Kabul: United Nations Assistance Mission in Afghanistan [UNAMA], 2012). Note also that civilians lose their protected status when they directly participate in hostilities (DPH) or are a part of "levée en masse," a term from the Third Geneva Convention describing a mass uprising of the civilian population. The author's position is that such civilians who lose protected status should not be tracked as civilian casualties, or alternately, be tracked separately.

termine the number of injured civilians; civilian deaths are easier to quantify, though easier does not mean simple.

Although it is not necessarily feasible to determine absolute numbers of civilian casualties overall, it is important to estimate the overall levels of casualties in the U.S. drone campaign as accurately as possible. While it might seem as if the U.S. government is best positioned to measure the impact of an activity, that may not be the case, as discussed below. Regardless of what the U.S. government knows, however, it does not routinely share this information with others. When data is shared, it typically comes in the form of quotes from U.S. military commanders and top government officials. While these figures point to very low numbers, they are not sufficient to generate an estimate.[3] Although President Obama recently promised greater transparency with regard to the U.S. drone campaign and its toll on civilians, actual numbers have yet to be released.[4]

The American government is not the only entity that can estimate the civilian impact of drone strikes. Several other organizations compile and track information on U.S. drone strikes in Pakistan and elsewhere, including the Bureau of Investigative Journalism (BIJ), a media organization headquartered in the United Kingdom, and the New America Foundation (NAF), a U.S. think tank.[5] A recent report by the UN special rapporteur, reflecting comments by the Pakistan government, provided another estimate of civilians killed in drone strikes in Pakistan from 2004 to 2013.[6]

NAF and BIJ share a similar methodology for aggregating numbers of casualties contained in news reporting. Both BIJ and NAF aim to increase

---

[3] It is not clear whether the United States has an existing database of civilian casualties from the drone campaign—akin to the practice of International Security Assistance Force (ISAF) and U.S. forces in Operation Enduring Freedom—that it could use to derive estimates.

[4] Peter Bergen and Jennifer Rowland, "Did Obama Keep His Drone Promises?," CNN News, 25 October 2013, http://www.cnn.com/2013/10/25/opinion/bergen-drone-promises/.

[5] BIJ's drone data is available at https://www.thebureauinvestigates.com/category/projects/drones/drones-graphs/; or see Daniel Rothenberg et al., *Reflections from the First Annual Conference on the Future of War* (Washington, DC: New America Foundation, 2015), https://www.newamerica.org/international-security/future-war/policy-papers/reflections-from-the-first-annual-conference-on-the-future-of-war-3/.

[6] Emmerson, *Promotion and Protection of Human Rights*.

transparency regarding the drone campaign by compiling data on drone strikes and presenting this data alongside the original reports. Both organizations reference a broad set of media dispatches, an approach that generates estimated ranges of casualties based on sometimes disparate reporting. For example, on 21 August 2012 in Shana Khora village, near Datta Khel in North Waziristan, Pakistan, witnesses saw four missiles from a drone impact a vehicle. BIJ estimated that there were between one and three civilian casualties from this strike, based on several available news reports. Overall casualty totals presented here include both the minimum and maximum values for reference; however, the detailed analysis in a later section uses the minimum number of casualties from these two sources.[7]

While these two organizations share the same general approach, they differ somewhat in the specific data sources. NAF relies on a group of newspapers it deems reputable.[8] Similarly, BIJ references a set of newspapers considered reliable, but it also considers additional sources, such as WikiLeaks and public interest lawsuit documentation, as well as its own field investigations in Pakistan.[9] The difference in approaches leads to some variance in civilian death estimates reported by the two organizations.

The UN special rapporteur report reflects information provided by the Pakistani government concerning its own estimates of civilian deaths from U.S. drone strikes. Similar to the BIJ and NAF estimates, the Pakistani

---

[7] It has been discussed elsewhere how U.S. official estimates for civilian casualties tend to be too low while media reports trend higher. See Larry Lewis and Sarah Holewinski, "Changing of the Guard: Civilian Protection for an Evolving Military," *PRISM* 4, no. 2 (2013): 57–66. The practice of using the minimum value is expected to help reduce the impact of this inflation factor observed in some reports.

[8] NAF uses the three major international wire services (Associated Press, Reuters, and Agence France-Presse), the leading Pakistani newspapers (*Dawn, Express Times, The News,* and *Daily Times*), leading Southeast Asian and Middle Eastern television networks (Geo TV and Al Jazeera), and Western media outlets with extensive reporting capabilities in Pakistan (CNN, *New York Times, Washington Post, LA Times*, BBC News, *The Guardian,* and *Daily Telegraph*). For more on the topic, see Ritika Singh, "A Meta-Study of Drone Strike Casualties," *Lawfare* (blog), 22 July 2013, https://www.lawfareblog.com/meta-study-of-drone-strike-casualties/.

[9] Founded in 2006 by Iceland native Julian Assange, WikiLeaks promotes itself as a nonprofit, journalistic organization that publishes leaked and often classified documents from anonymous sources.

government's estimates offer a range of values, with a minimum and an additional number of possible suspected noncombatants. The Pakistani government also stated that this number should be regarded as an underestimate of the true civilian toll due to challenges of access, investigation, and reporting in Pakistan's Federally Administered Tribal Areas (FATA).[10]

Estimates for total civilian casualties from U.S. drone strikes in Pakistan over the period (2004–13) from BIJ, NAF, and the UN are shown in figure 1.[11] The minimum numbers from each of these sources are used as the baseline values, with the maximum estimates also provided for reference. While the range is wide—spanning a minimum of 258 casualties for the minimum estimate of NAF to a maximum estimate of 951 for BIJ—the large disparity in these values is not hard to explain. First, while BIJ has a wide range of possible estimates, it regards the lower end of its estimates to be more likely to approach actual values. This assumption is consistent with analysis for Afghanistan that showed a propensity of some reporting of civilian casualty incidents to show inflated values for high-visibility incidents.[12]

In addition, while BIJ documents two categories of casualties (civilian and militant), NAF data covers three (civilian, militant, and unknown). Per international law, individuals of unknown status are to be treated as civilian. There were a minimum of 198 casualties in the NAF data identified as unknown; these should be included in civilian estimates barring evidence to the contrary, increasing the NAF minimum estimate to 456. The column

---

[10] The UN special rapporteur "was informed that the Government [of Pakistan] has been able to confirm that at least 400 civilians had been killed as a result of drone strikes, and that a further 200 individuals were regarded as probable noncombatants. Officials indicated that due to underreporting and obstacles to effective investigation on the ground these figures were likely to be under-estimates of the number of civilian deaths." *Statement of the Special Rapporteur following Meetings in Pakistan* (New York: United Nations, 14 March 2013), newsarchive.ohchr.org/EN/NewsEvents/Pages/DisplayNews.aspx?NewsID=13146&LangID=E.

[11] Numbers for BIJ were derived from BIJ's database, as provided by the author. Numbers for NAF were posted on their website for "Drone Wars Pakistan: Analysis," NAF, 31 December 2013.

[12] Lewis and Holewinski, "Changing of the Guard."

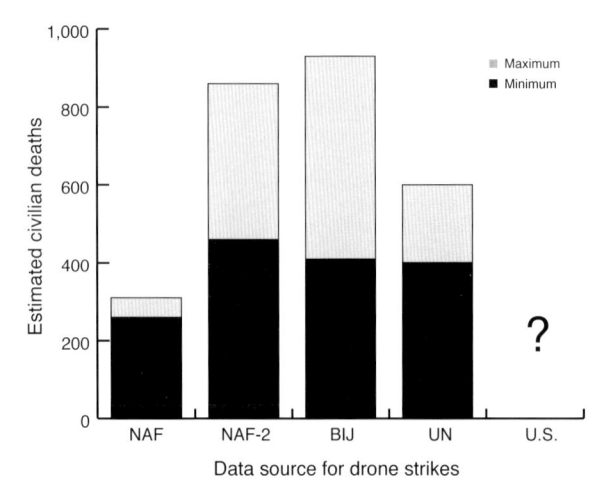

Figure 1. Disparate estimates for civilian deaths from drone strikes in Pakistan, 2004–13.

NAF-2 represents the number of civilian deaths plus those defined as "unknown."

Source: Bureau of Investigative Journalism and New America Foundation, "Drone Wars Pakistan: Analysis," accessed 31 December 2013.

labeled "NAF-2" in figure 1 reflects the addition of these casualties and appears more consistent with the other data sources.

Thus, the range of values for the UN estimate—between 400 and 600—could be viewed as a reasonable general range based on these considerations. Most notably, the UN source is independent of these two other data sets. It is possible, however, that these estimates are artificially low. For example, the Pakistani government acknowledged access limitations that could make its estimate lower than the actual civilian toll. Similarly, observers noted that some factors could lead the media to systematically underreport casualties, which would lower both NAF and BIJ estimates compared to actual values.[13] For example, news reports are affected by

---

[13] James Cavallaro, Stephan Sonnenberg, and Sarah Knuckey, *Living Under Drones: Death, Injury and Trauma to Civilians from US Drone Practices in Pakistan* (Stanford, CA: International Human Rights and Conflict Resolution Clinic, Stanford Law School and the Global Justice Clinic, NYU School of Law, 2012).

limited journalist access to the FATA areas, which can result in missing information available only in the local area, as well as reporters' reliance on intelligence channels that may be unaware of the actual extent of civilian harm, which will be discussed later in this section. Such systematic omissions could cause both BIJ and NAF estimates to be lower than actual civilian tolls, though BIJ estimates should be affected less by this factor due to its multisource data collection methodology.

While there is no official U.S. estimate to compare to these values, recent public U.S. government comments on the civilian toll of drone strikes suggest significantly smaller numbers of casualties compared to these other sources. For example:

- Former Deputy National Security Advisor John O. Brennan said in June 2011: "[Over the past 10 months,] there hasn't been a single collateral death because of the exceptional proficiency, precision of the capabilities we've been able to develop."[14]

- Senate Intelligence Committee Chair Senator Dianne Feinstein said in February 2013: numbers of civilian casualties each year for drone strikes overall, including both Pakistan and Yemen, have "typically been in the single digits."[15]

The process of counting casualties in conflict is not easy; it poses political as well as practical challenges. The fact that senior U.S. officials are repeatedly called to comment on civilian tolls from drone attacks highlights the political dimension of civilian casualties. The political pressure created by civilian casualties has been seen consistently in recent Coalition campaigns over the past several decades, reaching a high point in Afghanistan.

---

[14] Scott Shane, "CIA Is Disputed on Civilian Toll in Drone Strikes," *New York Times*, 11 August 2011, http://www.nytimes.com/2011/08/12/world/asia/12drones.html?_r=0.

[15] Lee Ferran, "Intel Chair: Civilian Drone Casualties in 'Single Digits' Year-to-Year," ABC News, 7 February 2013, http://abcnews.go.com/blogs/headlines/2013/02/intel-chair-civilian-drone-casualties-in-single-digits-year-to-year/.

CHAPTER 3

DISCREPANCIES IN CIVILIAN CASUALTY ESTIMATES

Some practical considerations make estimating civilian casualties difficult, which could explain why U.S. casualty estimates are lower than those from other sources mentioned previously. Three factors complicate the estimation process:

- **An irregular enemy.** The nature of the enemy in Pakistan, Afghanistan, and other locations can create challenges in positive identification (PID) of enemy combatants and discriminating between the civilian population and the enemy. These challenges result partly from the irregular nature of combatants (e.g., a lack of uniforms and standard equipment that limits the visual signature for PID, especially within an armed culture) and their practices of colocating with the local population, using noncombatants as human shields, and claiming—and creating—civilian casualties.

- **Misidentification.** These casualties occur when U.S. forces mistakenly identify civilians as enemy combatants. In engagements involving misidentification, because the casualties are believed to be combatants, they are not reported as civilians, and the reality is discovered later if, in fact, it is ever discovered by the United States.

- **Inaccurate assessments based on air surveillance.** The process of determining the effects of an engagement on the enemy and the surroundings—typically called battle damage assessments (BDA)—should include an assessment of any civilian toll from the engagement. However, given the irregular enemy and possibility of misidentification, this evaluation of civilian toll can be difficult to determine accurately. This is especially the case in situations when U.S. forces rely primarily on air surveillance for this assessment. Air assessments are likely the predominant method for BDA in the U.S. drone campaign.

These three factors are not specific to actions in Pakistan; rather, they are common to many counterterrorism scenarios where airpower is used. These factors are illustrated in the following two incidents from Afghanistan.[1]

DEH BALA AIRSTRIKE

On 5 July 2008, U.S. military forces targeted airstrikes against what they believed to be enemy combatants in a wooded area in Deh Bala District, Nangarhar Province. Shortly thereafter, the local population and media reports claimed that high levels of civilian casualties had resulted from the airstrikes.[2] A U.S. military spokesman immediately denied that there were civilian casualties:

> Whenever we do an airstrike the first thing they're going to cry is "airstrike killed civilians" when the missile actually struck mili-

---

[1] While processes and operating forces in Afghanistan can differ from those in drone operations in Pakistan, these operations share common elements. Also, Afghanistan holds the advantage of established reporting and investigative processes for civilian casualty incidents, facilitating more effective analysis.

[2] *Troops in Contact: Airstrikes and Civilian Deaths in Afghanistan* (New York: Human Rights Watch, 2008), http://www.hrw.org/reports/2008/afghanistan0908/.

tant extremists we were targeting in the first place. At this time we don't believe we've harmed anyone except for the combatants.[3]

Later, information surfaced that a group of civilians, walking from one village to another to participate in a wedding, was mistaken for combatants and engaged. Civilians were indeed killed by the U.S. airstrikes, but because the government believed them to be enemy combatants, the civilian toll was not acknowledged until locals found and identified the bodies. A subsequent U.S. inquiry confirmed that dozens of civilian casualties resulted from the airstrike.[4]

FARAH AIRSTRIKE

On 4 May 2009, Afghan security forces moved into the vicinity of Shewan, Farah Province, to confront a large group of Taliban that had moved into the area. The Afghan forces were ambushed, and a small contingent of U.S. advisors called for reinforcements from close air support and a nearby U.S. Marine quick reaction force (QRF). The Marines used airstrikes first to counter enemy attacks and then to target enemy combatants behind the line of battle, including several engagements at compounds.[5] U.S. forces conducted BDAs of the airstrikes that impacted close to them, but they did not inspect the strikes on the compounds in the village due to concerns over the safety of friendly forces.[6]

Reports of civilian casualties quickly emerged in the media, but the U.S. government initially denied the veracity of the reports. "It is certainly

---

[3] Amir Shah and Jason Straziuso, "Afghan Officials: U.S. Missiles Killed 27 Civilians," *USA Today*, 6 July 2008, http://usatoday30.usatoday.com/news/world/2008-07-06-1051356149_x .htm.

[4] See *Troops in Contact*.

[5] U.S. Central Command (CENTCOM) Public Affairs, *USCENTCOM's Unclassified Executive Summary: U.S. Central Command Investigation into Civilian Casualties in Farah Province, Afghanistan on 4 May 2009* (MacDill Air Force Base, Tampa, FL: CENTCOM Headquarters, 2009).

[6] In their role as a QRF, the U.S. Marines did not have the supplies to stay in the area for a prolonged period of time. Also, they believed the Afghan force was in the lead and would be responsible for necessary follow-on actions for the incident.

a technique of the Taliban and other insurgent groups to claim civilian casualties at every event," said International Security Assistance Force (ISAF) Commander General David D. McKiernan on 6 May.[7] On 15 May, U.S. Marine Corps Commandant General James T. Conway noted that, "We believe that there were families who were killed by the Taliban with grenades and rifle fire that were then paraded about and shown as casualties from the air strike."[8]

Shortly after the initial incident, U.S. Central Command (CENTCOM) assembled an investigation team and released an interim report on 15 May, confirming that dozens of civilian casualties had in fact occurred due to the U.S. airstrikes. The team's final report and unclassified summary were released shortly thereafter.[9]

These two examples illustrate the issues that develop when the United States conducts airstrikes that harm civilians and yet remain ignorant of the fact due to misidentification, inaccurate or missing BDAs, and the nature of the enemy. In Afghanistan, after a number of high-profile events, the American military addressed these challenges through additional guidance and procedures. For example, U.S. forces frequently conducted BDAs using ground forces when feasible, since ground-based assessments were far less likely to miss instances of civilian harm.

MISIDENTIFICATION: IMPACT ON CIVILIAN CASUALTIES AND ASSESSMENTS

Though all three factors described above—an irregular enemy, misidentifications, and inaccurate assessments of resultant harm—complicate the assessment of civilian casualties in operations, the issue of misidentification is particularly important. In previous analyses of civilian casualties in Afghanistan, the misidentification of civilians as the enemy was the basis for

---

[7] Donna Miles, "Joint Investigation Aims to Get to Bottom of Afghanistan Incident," DOD News, 6 May 2009, archive.defense.gov/news/newsarticle.aspx?id=54224.

[8] Gen James T. Conway, "Remarks by the Commandant of the Marine Corps" (speech, Center for Strategic and International Studies, Washington, DC, 15 May 2009).

[9] *U.S. CENTCOM's Unclassified Executive Summary.*

the majority of civilian casualty incidents and contributed to a lack of recognition of actual civilian tolls from operations.[10] Two issues contributed to this misidentification:

- **Misinterpretation of actions or characteristics.** In some cases, civilians were targeted because their behavior appeared threatening or their appearance marked them as enemy forces. For example, a number of Afghan civilians were killed because they appeared to be emplacing improvised explosive devices (IEDs). After they were engaged, the force discovered the Afghans were actually digging drainage ditches or doing other agricultural work. In other cases, individuals were targeted because they were suspected of carrying weapons. After the engagement, it came to light that they were holding farming tools or other large objects. And while being armed is no guarantee of nefarious activity, many civilians—and even Afghan forces—are accidentally killed for this reason.

- **Guilt by association.** In some incidents, enemy forces were located in close proximity to civilians who were not directly participating in hostilities. However, when U.S. forces engaged the enemy forces, the nearby civilians were also believed to be enemy and killed or wounded.

A single incident in Uruzgan Province in Afghanistan from 2010 illustrates both of these considerations. U.S. forces believed two civilian vehicles carried Taliban fighters with the intent to attack a U.S. Special Operations Forces (SOF) element in the area. A U.S. Predator drone crew misidentified these vehicles as enemy forces through misinterpretation of their actions. When a third vehicle joined the other two, the Predator crew considered the third vehicle to be enemy due to guilt by association. Based on this mis-

---

[10] Larry Lewis, *Reducing and Mitigating Civilian Casualties: Enduring Lessons* (Suffolk, VA: Joint Civilian Casualty Study by Joint and Coalition Operational Analysis [JCOA], 2013).

identification, a U.S. helicopter later engaged the three vehicles, resulting in dozens of civilian casualties.[11]

These mechanisms of misidentification can also impact assessments of civilian casualties. However, just because civilians are colocated with enemy forces does not mean that the engagement is not permissible. Under U.S. and international humanitarian laws, such as the Geneva Conventions, force may be used against an enemy as long as the harm to civilians is not excessive relative to the gained advantage from the operation.[12] These civilian tolls should be properly acknowledged in follow-on reporting and assessments. In Afghanistan, the U.S. military initially counted misidentified civilians as enemy personnel. This error was recognized in a subsequent assessment process that examined underlying assumptions; the misidentified individuals were then counted as civilian. This process is illustrated by the Deh Bala and Farah airstrikes discussed earlier.

A number of independent reports describe drone strikes against buildings, convoys of vehicles, and groups of individuals. In these cases, individuals should not be counted as enemy personnel simply based on proximity to a known target. One overarching principle that should inform both engagements and assessments is a tenet of IHL used by international military forces in Afghanistan as well as by the UN: "In case of doubt whether a person is a civilian, that person shall be considered to be a civilian."[13] Media reports suggest that this has not been the approach guiding official U.S. assessments of civilian casualties in the drone campaign, with such descriptions as: the United States "counts all military-age males in a strike zone as combatants . . . unless there is explicit intelligence posthumously

---

[11] This incident is discussed in more detail in chapter 15.

[12] The U.S. military summarized customary international humanitarian law in this respect: ". . . loss of life and damage to property must not be out of proportion to the military advantage to be gained." *The Law of Land Warfare*, FM 27-10 (Washington, DC: 1956, modified by Change No. 1, 1976 rev.).

[13] See *Protocol Additional to the Geneva Conventions of 12 August 1949, and relating to the Protection of Victims of International Armed Conflicts (Protocol I)* (New York: 1977), hereafter *Protocol Additional to the Geneva Conventions of 12 August 1949*, https://www.icrc.org/ihl/WebART/470-750064?OpenDocument.

proving them innocent."[14] If true, this approach is inconsistent with both international law and U.S. military practice in Afghanistan, and would lead to an inaccurate picture of the civilian toll from those strikes.[15]

The U.S. drone campaign is characterized by airborne target identification and BDA. These factors create opportunities for misidentification in irregular warfare, and increase the likelihood that civilians, including those misidentified as enemy, are not discovered by the American government. Thus, it is likely that the United States does not have a true picture of the actual scale of civilian harm from its drone campaign. Regarding operations in Pakistan and Yemen, the presence of civilian casualties reported in the media has frequently been denied. This resembles the situation in Afghanistan prior to mid-2009, where U.S. and international military commanders were frequently confronted by reports of civilian casualties that differed from their own initial reports, as the above examples illustrate.

It is important to note that the challenge of recognizing civilian harm from drone strikes in Pakistan can be even more challenging than it was in Afghanistan, due to the reduced number of U.S. boots on the ground and limited communication with local forces and communities. That said, the United States could find ways to compensate for these additional challenges, for example, by partnering with third-party organizations with a presence on the ground or through increased reliance on leveraging human intelligence (HUMINT) to cue other intelligence sources to enhance BDA.

In both Afghanistan and the current drone campaign in Pakistan, the stated desire of the United States to minimize civilian harm was evidenced by such statements as: "We've done everything possible . . . to reduce any risk to that civilian population."[16] However, the military's ability to do ev-

---

[14] Jo Becker and Scott Shane, "Secret 'Kill List' Proves a Test of Obama's Principles and Will," *New York Times*, 29 May 2012, http://www.nytimes.com/2012/05/29/world/obamas-leadership-in-war-on-al-qaeda.html?pagewanted=all.

[15] While the legal framework for counterterrorism (CT) strikes in Pakistan is in debate, the international norm of the default status of individuals to be civilians in case of uncertainty would appear to be valuable to preserve. Erosion of this norm could eliminate the requirement for discrimination of targets in the context of CT operations.

[16] Dilanian, "Brennan Defends U.S. Drone Attacks Despite Risks to Civilians."

erything possible to avert civilian harm is limited by its capacity to consistently recognize instances of civilian harm. If the problem of civilian harm is not recognized and well understood, then the actual scale of the damage done will be misunderstood and measures will not be put in place to address it effectively. Thus, an assessment process to quantify levels of civilian harm is needed to ensure that U.S. efforts truly minimize civilian harm.

CHAPTER 4

PLATFORM PRECISION OR COMPREHENSIVE PROCESS?

Public statements defending drone use often comment on the precise nature of a platform with the ability to engage intended targets without causing civilian harm. For example, one statement in the public debate on drones declares: "Where civilian casualties cannot be avoided, they must be minimized. This is what drone strikes do."[1] This viewpoint suggests that using the drone platform to engage the enemy constitutes all the steps America needs to minimize civilian harm. But, in fact, this statement is incorrect for several reasons: it mistakes platform precision for a comprehensive process that minimizes civilian casualties; and it is contradicted by operational data.

PRECISION VERSUS PROCESS

Although drones offer desirable capabilities, such as precision weapons, persistence, and full-motion capabilities for targeting and screening of collateral damage, these technical elements alone do not necessarily translate

---

[1] Avery Plaw, "Drone Strikes Save Lives, American and Other," *New York Times*, 14 November 2012, http://www.nytimes.com/roomfordebate/2012/09/25/do-drone-attacks-do-more-harm-than-good/drone-strikes-save-lives-american-and-other.

to surgical precision and the minimization of civilian casualties. Other factors also influence the likelihood of civilian casualties.[2]

The importance of the overall process—the collective impact of the different factors shaping individual engagement decisions on civilian casualty reduction—was discussed previously in the 2010 Joint Civilian Casualty Study (JCCS), which examined operations in Afghanistan for the U.S. military. A comprehensive approach to civilian casualty reduction and mitigation was envisioned, including a number of different steps in the civilian casualty "lifecycle" (figure 2):

- **Prepare:** doctrine, professional military education, predeployment training and equipping, exercises, training, and adaptation

- **Plan:** mission planning, rehearsals, intelligence, pattern of life, and other information, as well as shaping the environment

- **Employ:** the use of force, tactical alternatives, the rules of engagement, and tactical directives

- **Assess:** battle damage assessments, data collection, and data sharing

- **Respond:** medical response, key leader engagement, media engagement, providing compensation, other information activities

- **Learn:** reporting, data management, data analysis, after action reports, investigations, and capturing and disseminating lessons[3]

As the lifecycle illustrates, minimizing civilian casualties is less a matter of platform or ordnance selection than using an approach that considers factors leading to civilian casualties and then effectively takes them into account. In particular, the importance of "learning" to minimizing civilian

---

[2] The prolonged use of the AGM-114 Hellfire air-to-surface missile (ASM) in drone strikes shows both adaptability with existing capabilities—using a missile originally designed for helicopters to attack tanks—and the occasionally slow development of military capabilities to reduce civilian casualties. See Lewis, *Reducing and Mitigating Civilian Casualties*.

[3] Ibid.

Figure 2. Comprehensive process for reducing and mitigating civilian harm

Prepare
• Predeployment training
• Doctrine
• Education

Learn
• After action reports
• Investigations
• Dissemination of lessons to force
• Feedback to institutions

Plan
• Mission planning
• Coordination
• Rehearsals
• Intelligence and information

Respond
• Medical
• Key leader engagement
• Media engagement
• Amends and apologies

Employ
• Application of guidance for the use of force
• Tactical patience
• Tactical alternatives

Assess
• Hold the ground
• Battle damage assessment
• Establish command and control
• Reporting

casualties during operations is missed when attention is instead focused on the *platform* rather than the *process*. The significance of learning to reduce civilian casualties can be seen in the decrease of Iraqi casualties from escalation of force incidents in 2005 and 2006.[4] Lieutenant General Peter W. Chiarelli, the Multinational Corps–Iraq commander, helped focus U.S. forces on primary causal factors to learn from past incidents and not repeat the same mistakes. Civilian casualties dropped significantly as a result.[5] Similarly, but on a larger scale, commanders in Afghanistan began tracking civilian casualties for all types of Coalition-caused incidents, using analysis to identify causal factors and to reshape their guidance. For example, analysis of Coalition air operations, documented in the JCCS report, led to changes in the 2010 Commander, International Security Assistance Force

---

[4] Thom Shanker, "New Guidelines Aim to Reduce Civilian Deaths in Iraq," *New York Times*, 21 June 2006, http://www.nytimes.com/2006/06/21/world/middleeast/21cnd-casualties.html?pagewanted=print&_r=0; and Nancy A. Youssef, "U.S. Working to Reduce Civilian Casualties in Iraq," McClatchy DC, 21 June 2006, http://www.mcclatchydc.com/latest-news/article24455704.html.

[5] Lewis, *Reducing and Mitigating Civilian Casualties*.

(COMISAF) Tactical Directive, which then was seen to reduce the lethality of civilian casualty incidents.[6] Overall, international observers, such as the UN, have acknowledged that the United States has made significant progress in reducing Coalition-caused civilian casualties in Afghanistan.

While the progress reducing civilian casualties in Afghanistan shows what is possible, to date, the changes put into place remained focused largely on supporting operations there. In 2013, however, the U.S. military proactively began to focus on institutionalizing key enduring lessons for the future force. Though, sharing lessons in different operations and among allied countries is less apparent. For example, key lessons and best practices from Afghanistan were not known to North Atlantic Treat Organization (NATO) forces in Libya, forcing discovery learning on the ground. In addition, it is unclear whether lessons from Afghanistan have been applied to the U.S. drone campaign in Pakistan and elsewhere.

OPERATIONAL DATA:
DRONES MORE LIKELY TO CAUSE CIVILIAN HARM

Operational data confirms that reducing civilian casualties depends on the entire engagement process, including planning and training considerations, not simply on the characteristics of the weapon platform. Analysis of data from Afghanistan showed that several forms of attack, including engagements by manned air platforms, were less likely to cause civilian casualties than drone strikes, highlighting the fact that platform characteristics alone are not the driver of a decreased likelihood of civilian casualties.[7]

The discussion of process shows how analysis and assessment can provide insight into trends and highlight the root causes of civilian casualty incidents. Later, we will present a model for an assessment process, including root cause analysis of a real-world civilian casualty incident, and outline how the process can be used to better minimize civilian harm.

---

[6] Ibid.

[7] *Drone Strikes: Civilian Casualty Considerations* (Suffolk, VA: JCOA, 2013).

CHAPTER 5

THE DRONE CAMPAIGN AND CIVILIAN HARM

Congress has debated whether the drone campaign should be shifted entirely to the U.S military or whether it should continue to be conducted, in part, by another element of the U.S. government.[1] Many considerations must go into this decision, particularly the ability of the organization leading the drone campaign to minimize civilian harm. One question that could be asked is: how well is the current drone campaign in Pakistan doing in minimizing civilian harm?

Table 1 shows the total number of drone strikes (engagements), the numbers of civilians killed, and the number of strikes that resulted in civilian casualties (civilian casualty incidents) from two data sources—BIJ and NAF. For NAF, two sets of numbers are provided: one includes only confirmed civilians (NAF); and one includes unknown casualties treated as dictated per international law (NAF-2).

Table 1 indicates that, on average, one civilian has been killed by each drone strike since the inception of the campaign. BIJ and the adjusted NAF databases report the same rates of strikes causing civilian casualties, with one in every five strikes killing civilians. For these strikes, between five and six civilians are killed on average (figure 3).

---

[1] Dylan Matthews, "Everything You Need to Know about the Drone Debate, in One FAQ," *Washington Post* (Wonkblog), 8 March 2013, http://www.washingtonpost.com/news/wonkblog/wp/2013/03/08/everything-you-need-to-know-about-the-drone-debate-in-one-faq/.

Table 1. Overall statistics for drone strikes in Pakistan, 2004–13

	BIJ	NAF	NAF-2
Overall drone strikes (engagements)	383	370	370
Civilians killed (CIV K)	416	258	456
Strikes where CIV were killed (CIV K incidents)	75	32	75
Average CIV K per engagement	1.1	0.7	1.2
CIV K incidents per engagement (percent)	20	9	20
CIV K per CIV K incident	5.5	8.1	6.1

Source: Bureau of Investigative Journalism and New America Foundation, data for 2004–13.

Of course, the U.S. drone campaign in Pakistan was not consistent in the number of strikes over time. It was carried out at a low initial rate, increased significantly in 2008, peaked in 2010, and then tapered off gradually after 2010 (figure 3). Accordingly, it is instructive to examine characteristics from each period.

Pre-2008 strikes were characterized by very few attacks, with a high likelihood of civilian casualties per engagement (64 percent and 70 percent for BIJ and the adjusted NAF, respectively) and a high average civilian toll for incidents where civilians were killed. Starting in 2008–9, the number of strikes increased significantly, and the rate of civilian deaths per engagement dropped significantly (34 percent for BIJ compared to 32 percent for the adjusted NAF). Starting in 2010, the rate of engagements causing civilian deaths drops to approximately 13 percent for BIJ and 14 percent for the adjusted NAF. During this time, an average of one civilian was killed for about every two drone strikes, and the civilian toll for incidents that caused civilian casualties was about four per incident.

Figure 3. Number of drone strikes in Pakistan per year

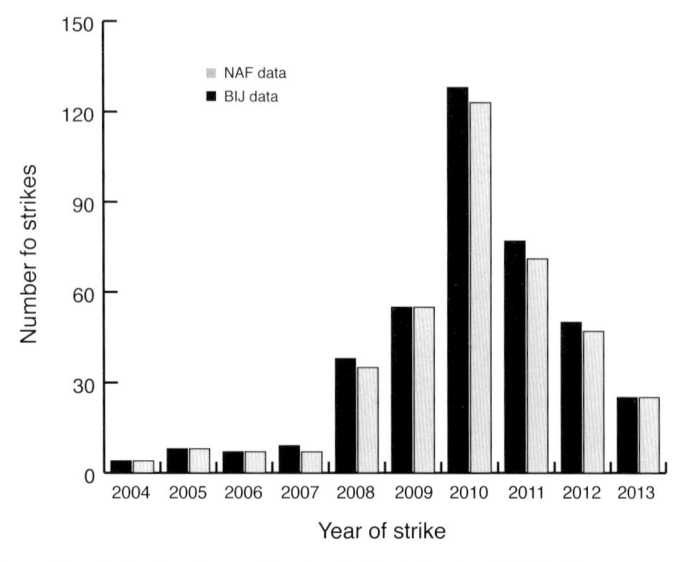

Source: Bureau of Investigative Journalism and New America Foundation, data for 2004–13.

Figure 4. Percent of operations resulting in civilian deaths per year

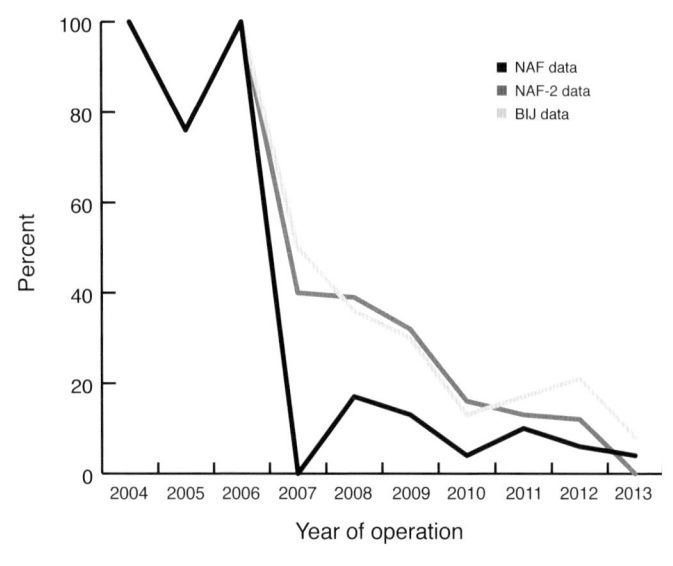

Source: Bureau of Investigative Journalism and New America Foundation, data for 2004–13.

26 | Drone Strikes in Pakistan

For our discussion, it is also helpful to plot key metrics over time. The percent of operations resulting in civilian deaths offers a useful metric showing the average likelihood that a drone strike will result in civilian deaths (figure 4).

As seen above, the rate of civilian deaths per drone strike decreased over time, with rates generally less than 20 percent starting in 2010 for both BIJ and the adjusted NAF, and less than 10 percent for 2013. Note the generally close agreement between BIJ and the adjusted NAF where "unknown" status casualties are included, treating them as civilian as prescribed in international law.[2] Collectively, U.S. drone strikes have become less likely to cause civilian deaths over time.

However, it appears that there is still room for improvement. These civilian casualty rates for Pakistan are significantly higher than those for drone and overall CT operations in Afghanistan conducted by U.S. and international forces. While rates for the two countries are not necessarily directly comparable, the operations in Afghanistan illustrate that lower rates can be achieved during CT operations in general.

ASSESSMENT:
A KEY ELEMENT IN DEMONSTRATING CONCERN

Overall, it is both possible and worthwhile for the United States to conduct an independent assessment of civilian casualties resulting from drone strikes in Afghanistan, Pakistan, and Yemen. The results of this assessment can both inform refinements that reduce civilian harm in the future, as well as provide estimates that allow the legislative and executive branches to

---

[2] Media and other reporting on civilian casualties using NAF tends to neglect the "unknown" category of casualties. For example, CNN reported that "today, for the first time, the estimated civilian death rate is at or close to zero" when, in fact, the adjusted rate was higher than in the previous two years. Peter Bergen and Jennifer Rowland, "Civilian Casualties Plummet in Drone Strikes," CNN News, 14 July 2012, http://www.cnn.com/2012/07/13/opinion/bergen-civilian-casualties/. In another example, Brookings uses NAF as its data source on Pakistan drone strikes in its Afghanistan indicators publication; and these totals include only confirmed civilians. Ian S. Livingston and Michael O'Hanlon, *Afghanistan Index: Also Including Selected Data on Pakistan* (Washington, DC: Brookings, 2013), http://www.brookings.edu/wp-content/uploads/2016/07/index20130827.pdf

support greater transparency and enable improved oversight of these operations. In addition, assessment ensures that official U.S. statements are in line with operational realities, helping to guard the reputation of the nation.

DEMONSTRATING CONCERN THROUGH CONSEQUENCE MANAGEMENT

In addition to assessment, a process could be put into place to respond to U.S.-caused civilian casualties with consequence management actions, including restitution, when they occur from such strikes. This practice could use existing programs in a way similar to U.S. measures taken in Afghanistan, and those consistent with recent legislation granting authority for *ex gratia* payments for civilian casualties during military operations.

In Afghanistan since 2009, U.S. forces placed attention not only on reducing civilian casualties, but also responding to them in a moral and operationally effective way when they occurred. When civilian harm occurred as an unintentional result of U.S. operations, the government typically offered an apology. Providing monetary payments or other compensation, typically offered without admission of legal culpability, assisted families dealing with the financial and emotional components of their loss and reinforced the country's reputation as a nation that respects and upholds the lives of civilians.[3] This process also yielded operational benefits for U.S. forces by way of increased freedom of movement and willing support from the population.[4]

In Pakistan, the United States offers a program that aids communities impacted by conflict. The Conflict Victims Support Project provides rehabilitation and livelihood assistance among other things.[5] However, U.S. drone strike victims and their families are not currently covered by this

---

[3] This is consistent with a number of operations over the past century, where the United States offered compensation or aid to mitigate the impact of its actions.

[4] Lewis, *Reducing and Mitigating Civilian Casualties*.

[5] For more information on this U.S. Agency for International Development (USAID) program, see https://scms.usaid.gov/sites/default/files/documents/1871/CVSP.pdf.

program. Similarly, U.S. aid is not available for conflict victims in Yemen or other locations where drones also operate.[6] Such an effort could be conducted in partnership with other organizations to avoid a direct U.S. role, for example, the government of Pakistan or nongovernmental organizations. Besides offering U.S. concern for civilians, such a program would also support U.S. efforts to accurately identify and estimate civilian casualties. Further, adversaries that routinely exploit U.S.-caused casualties to discredit or tarnish America's reputation, and use this issue to solicit support for their cause, would find their case weakened if the nation could provide restitution for civilian harm it caused.

Collectively, an assessment process for civilian harm, coupled with measures to address such harm when it is caused, would further demonstrate U.S. concern for civilians caught in the middle of conflict, while also reducing grievances that exacerbate threats to the United States in the longer term. These initiatives could help the nation demonstrate its stated commitment to the responsible use of force and to minimize civilian harm in its operations.

---

[6] While an aid program does not exist, the federal government did budget $64 million for fiscal year 2014 in what it calls "counterterrorism security assistance." See Greg Miller, "Yemini Victims of U.S. Military Drone Strike Get More Than $1 Million in Compensation," *Washington Post*, 18 August 2014, https://www.washingtonpost.com/world/national-security/yemeni-victims-of-us-military-drone-strike-get-more-than-1million-in-compensation/2014/08/18/670926f0-26e4-11e4-8593-da634b334390_story.html. See also Scott Shane, "Families of Drone Strike Victims in Yemen File Suit in Washington," *New York Times*, 8 June 2015, http://www.nytimes.com/2015/06/09/world/middleeast/families-of-drone-strike-victims-in-yemen-file-suit-in-washington.html?_r=0.

CHAPTER 6

CONCLUSIONS AND RECOMMENDATIONS

The public debate on the drone campaign has unfortunately focused on the platform rather than the key issues at play: the legality of the use of force outside of declared theaters of conflict, and the ability of the United States to limit the civilian toll from the use of force in those operations.[1] These two issues are distinct, however, and can be debated and addressed separately: first, the legal issue with a long timeline to resolve; and second, addressed here, a policy issue within the U.S. government's ability to act quickly. Efforts to limit the civilian toll of the U.S. drone campaign do not need to be delayed simply because of disputes over other aspects of the drone campaign. Precisely, the government could immediately undertake an independent assessment of its drone operations in Pakistan, including a specific priority to analyze civilian casualties, to promote civilian harm response, and to address challenges in the targeting process that may put civilians in unnecessary danger.

Such a move could have several benefits. Working to reduce civilian casualties in U.S. operations also decreases the extent of radicalization and

---

[1] While there are other issues that can be debated, such as the appropriate role of automated systems in warfare, these two issues seem to be the primary concerns in the current public debate on the U.S. drone campaign. In her blog, Charli Carpenter discusses additional considerations for decomposing the key issues under debate with regard to the use of drones. See "Parsing the Anti-Drone Debate," *Duck of Minerva* (blog), 12 November 2013, http://duckofminerva.com/2013/11/parsing-the-anti-drone-debate.html.

support of threats to the nation and its interests. At the same time, operations with lower levels of civilian casualties would support freedom of action in future operations, promoting the ability to respond to imminent threats over the long term. Such an effort could also help fulfill the U.S. commitment to minimize civilian harm as a result of its operations. This alignment of practice and principle reinforces America's moral authority, enabling its continued global leadership.

RECOMMENDATIONS

First, conduct an independent U.S. government review of civilian harm in drone strikes, including a revised estimate of civilian casualties. U.S. drone strikes, past and present, should be analyzed to identify both levels and root causes of civilian harm, particularly:

- **Determine numbers and trends:** the government should review possible civilian harm in cases where credible evidence of such harm exists. This review might include information sources available in the states, such as video feeds and available intelligence, as well as those provided by other government and international organizations concerning these incidents. Such reviews were done for some instances in Afghanistan; with modifications, a similar process is feasible for U.S. drone strikes in Pakistan. These numbers can then be used to determine trends, similar to what was done here, providing a baseline assessment of possible progress and highlighting areas of particular concern. To be consistent with international law, data should include both confirmed civilians and casualties whose status has not been conclusively determined.

- **Assess root causes of civilian harm:** a key element to reducing civilian casualties in Afghanistan was the analysis of individual incidents to determine causal factors. When these factors were considered collectively, they focused efforts for reducing civilian harm to areas that were most productive. This process, which was conducted for ISAF and Operation Enduring Freedom commands, could easily be replicated for the U.S. drone campaign. A review similar to a safety investigation by the U.S. military could be conducted for instances of possible civilian harm to determine the likelihood of such harm and the causal factors for the incident. Periodic reviews would then consider these causal factors and identify ways to systematically address them.

Second, make civilian harm a component of congressional oversight for drone operations. Congress plays a role in shaping and validating U.S. policy through its oversight activities. For any operation that involves the use of force, the issue of civilian casualties remains a critical component to consider, as recent history has shown that civilian harm can derail a military campaign or undermine U.S. objectives if not handled effectively.[2] The importance of this issue is only likely to increase due to the growing transparency of overseas operations, greater scrutiny by external actors, higher expectations for the United States and its conduct of operations, and exploitation of civilian casualties by others to undermine the nation and oppose its interests. These realities point to the need for Congress to monitor civilian harm in periodic operational assessments, along with other appropriate indicators of mission effectiveness.

Third, apply best practices for civilian casualty reduction to the U.S. drone campaign. ISAF and U.S. operations in Afghanistan have made significant progress in reducing civilian casualties while maintaining mission effectiveness, including development of revised training, doctrine, tracking and analysis systems, weapons, and formalized responses to civilian harm. A number of these best practices and lessons could be applied to the drone activities outside Afghanistan. U.S. government elements conducting this campaign, including leaders and those responsible for executing operations, should seek out these lessons. The U.S. military's J7 Joint Force De-

---

[2] Besides the issue of civilian casualties becoming toxic in Afghanistan, civilian casualties also harmed CT operations in Iraq in 2009 and drove the development of restrictive policy guidance for those operations in Pakistan and Yemen, such as those in the Presidential Policy Guidance given on 23 May 2013. See White House, Office of the Press Secretary, "Fact Sheet: The President's May 23 Speech on Counterterrorism," https://www.whitehouse.gov/the-press-office/2013/05/23/fact-sheet-president-s-may-23-speech-counterterrorism.

velopment, which recently led the Joint Staff Civilian Casualties (CIVCAS) Working Group, is a good source for these lessons.[3]

From their perspective, it is critical to resolve previously identified challenges associated with drones and civilian casualties, as observed in Afghanistan operations. Applying lessons learned to the drone campaign requires a focus on previous analysis of U.S. and Coalition drone/UAV operations in Afghanistan, but also examining to what extent the same lessons and contributing factors apply. These specific areas should be addressed to ensure that the U.S. military can minimize the risk of civilian casualties from drone strikes, since civilian casualty rates for drone strikes were 10 times higher than that for manned aircraft and other types of engagements.[4]

Fourth, expand U.S. programs for victims of conflict to include the drone campaign. For example, the government routinely offered monetary payments, livelihood aid, and medical assistance to civilians harmed by combat operations in Iraq and Afghanistan. Providing similar assistance to the families of drone strike victims would provide humanitarian benefits to those suffering from U.S. actions, reinforcing the reputation of the United States. A policy of assistance would also undercut the development of grievances and exploitation of U.S.-caused casualties in war, providing longer term security benefits.

---

[3] For more information about J7 Joint Force Development, see http://www.jcs.mil/Directorates/J7%7CJointForceDevelopment.aspx.

[4] Sarah Holewinski and Larry Lewis, "Five Ways Obama Can Fix Drones Right Now," Defense One, 6 November 2013, http://www.defenseone.com/ideas/2013/11/five-ways-obama-can-fix-drones-right-now/73304.

PART II

THE FUTURE OF DRONE STRIKES

A FRAMEWORK FOR
ANALYZING POLICY OPTIONS
BY DIANE M. VAVRICHEK

CHAPTER 7

INTRODUCTION

Since the terrorist attacks on 11 September 2001, a steady stream of terrorist plots have materialized against U.S. targets and citizens.[1] The public will likely never know how many attacks have been thwarted by law enforcement, the military, and the intelligence agencies, but the would-be "Christmas Day" and "Times Square" bombers and the Boston Marathon bombings make clear the gravity of the threat of terrorism.[2] The U.S. government has been using a wide variety of methods in its fight to prevent terrorist attacks. One of those includes the use of armed drones.

UAVs, known colloquially as drones, have been used to carry out targeted killings since at least 2002, when the United States launched an

---

[1] One report documented 60 terrorist plots against the United States from 11 September 2001 through 2013. See Jessica Zuckerman, Steven P. Bucci, and James Jay Carafano, *60 Terrorist Plots Since 9/11: Continued Lessons in Domestic Counterterrorism*, Special Report #137 on Terrorism (Washington, DC: Heritage Foundation, 2013), http://www.heritage.org/research/reports/2013/07/60-terrorist-plots-since-911-continued-lessons-in-domestic-counterterrorism.

[2] The "Christmas Day" bomber, Umar Farouk Abdulmutallab, attempted to detonate a bomb hidden in his underwear during a flight to the United States in 2009. The "Times Square" bomber, Faisal Shahzad, attempted to detonate a car bomb in New York City's Times Square in 2010. Both bombs failed to detonate, and the bombers were subsequently sentenced by U.S. courts to life in prison. See, for example, Zuckerman, Bucci, and Carafano, *60 Terrorist Plots Since 9/11*. Two brothers, Dzhokhar and Tamerlan Tsarnaev, set off bombs that exploded near the finish line of the 2013 Boston Marathon. Tamerlan Tsarnaev was killed in a subsequent confrontation with police, and Dzhokhar was captured and convicted for the crime.

attack in Afghanistan against suspected al-Qaeda members.[3] Since then, targeted killings via drone strikes have become an integral piece of the U.S. campaigns Operation Enduring Freedom in Afghanistan, as well as against al-Qaeda and its associated terrorist groups, reportedly in Pakistan, Yemen, and Somalia.

The U.S. government claims that these strikes are effective. Indeed, in a 2013 speech, President Obama stated:

> Dozens of highly skilled al Qaeda commanders, trainers, bomb makers and operatives have been taken off the battlefield. Plots have been disrupted that would have targeted international aviation, U.S. transit systems, European cities and our troops in Afghanistan. Simply put, these strikes have saved lives.[4]

Drones have proven to be almost an ideal tool for the United States in many respects. They can execute kinetic attacks while limiting collateral damage due to their accuracy and ability to surveil targets for hours to determine the presence of other people in the area before striking.[5] They do not risk the life of a pilot and may be less expensive to operate than manned aircraft. Furthermore, drones enable targeted killing operations in locations where the United States does not have a presence on the ground.

Moreover, with the problems surrounding Guantánamo Bay Naval Base in Cuba, no other readily apparent long-term detention options available for dealing with captured terrorists, and the planned withdrawal of U.S. forces from large-scale combat operations, some proponents argue

---

[3] Secretary of Defense Donald H. Rumsfeld, DOD News Briefing, 12 February 2002, https://www.fas.org/irp/news/2002/02/dod021202.html, hereafter Rumsfeld Briefing.

[4] Obama, "Remarks by the President at the National Defense University."

[5] Despite these advantages, however, U.S. claims that drones are "surgical" in their ability to avoid civilian casualties seem not to be supported by operational data. The analysis of strikes in Afghanistan mentioned in part 1 showed that drone strikes were 10 times more likely to cause civilian casualties than strikes by manned aircraft.

that drone strikes are becoming one of few remaining tools left to use in the U.S. counterterrorism (CT) mission.[6]

Yet the practice of using drones in targeted killing operations has become highly controversial, both within the United States and in the international community, for numerous reasons:

- The United States has only vaguely publicly explained its targeting processes and standards amid growing public pressure for increased transparency and accountability. The UN High Commissioner for Human Rights went so far as to note that, "[t]he current lack of transparency surrounding [the use of drone strikes] creates an accountability vacuum."[7]

- There is controversy in the international community as to whether the practice is in accordance with international law, as detailed below.

- There is controversy within the United States around the authorization of the practice under domestic law, including debate over the 2001 Authorization for the Use of Military Force (AUMF) and the role of the War Powers Resolution.[8]

- Highly publicized incidents document that innocent people have been mistakenly killed.[9]

---

[6] See, for example, Max Fisher, "The Best Case for Drones I've Heard Yet," *Washington Post*, 30 December 2013, http://www.washingtonpost.com/blogs/worldviews/wp/2013/12/30/the-best-case-for-drones-ive-ever-heard/.

[7] Navanethem Pillay, "Pillay Briefs Security Council on Protection of Civilians on Anniversary of Baghdad Bombing" (statement, Security Council, New York, 19 August 2013), http://www.ohchr.org/EN/NewsEvents/Pages/DisplayNews.aspx?NewsID=13642&LangID=E.

[8] Harold Hongju Koh, "Libya and War Powers" (testimony, U.S. Senate Committee on Foreign Relations Washington, DC, 28 June 2011), http://www.state.gov/s/l/releases/remarks/167250.htm. These important debates fall outside the scope of this report.

[9] For example, Hakim Almasmari, "Yemen Says U.S. Drone Struck a Wedding Convoy, Killing 14," CNN News, 13 December 2013, http://www.cnn.com/2013/12/12/world/meast/yemen-u-s-drone-wedding/.

- The U.S. government targeted and killed an American citizen.[10]

- In addition to targeting identified individuals, the United States engages in "signature strikes" in which reportedly "a strike is authorized based on patterns of behavior in an area but where the identity of those who could be killed is not known."[11]

- Other countries and their militaries are increasingly investing in drones; there is growing concern that the United States is not setting a standard for the use of drone strikes that commands sufficient restraint and transparency.

Because of these controversies, numerous proposals have been made for changes to U.S. drone strike policy from both within and outside of the government. Notably, in May 2013, President Obama issued guidance to implement certain changes, although that implementation has been at least partially blocked by Congress.[12] We propose here a general framework for evaluating proposed policy options for drone strike operations, with a focus on tactical military effectiveness and the public perception of legitimacy, and how these two factors can influence the net effectiveness of the operations within the larger CT context. In its analytic approach to evaluating tactical effects and incorporating these with the implications of public

---

[10] The significant issues and controversies around targeting American citizens also fall outside the scope of this work.

[11] "Letter from Congressmen John Conyers Jr., Jerrold Nadler, and Robert C. Scott to Attorney General Eric H. Holder Jr.," 4 December 2012, http://www.propublica.org/documents/item/605032-conyers-nadler-scott121204.

[12] "Background Briefing by Senior Administration Officials on the President's Speech on Counterterrorism" (briefing, White House, Office of the Press Secretary, 23 May 2013), www.whitehouse.gov/the-press-office/2013/05/23/background-briefing-senior-administration-officials-presidents-speech-co, hereafter "Background Briefing"; and Eric Schmitt, "Congress Restricts Drones Program Shift," *New York Times*, 16 January 2014, http://www.nytimes.com/2014/01/17/us/politics/congress-restricts-drones-program-shift.html?_r=0; and John T. Bennett, "McCain Vows New Fight over Control of U.S. Armed Drone Program," Defense News, 19 February 2014.

perceptions, this framework addresses a gap in the current dialogue over drone strikes.

The next section of this work briefly outlines the framework and provides an overview of President Obama's guidance and several other policy options. After that, methods for evaluating the effects of proposed changes to drone strike policy on the framework are presented in three areas: military effectiveness, legitimacy issues, and how these topics contribute to the net effectiveness of a policy option on the broader goals of U.S. security and counterterrorism. Before moving on, however, this section closes with a few words on applying the framework in more general settings.

APPLYING THE FRAMEWORK TO TARGETING IN GENERAL

In some sense, there is nothing special about the role that drones, as a weapons platform, play in this discussion; the framework below could be applied equally well in the broader context of remote targeting and even targeting in general. However, a broader application of this analysis to targeting in general should be performed with caution.

One reason to exercise caution is that, while this work considers drone strikes within the context of the CT campaign (and, to a lesser extent, that of the "hot" battlefield of Afghanistan), targeting operations could potentially take place within vastly different settings and as components of completely different missions. Any such situation should be considered individually and the framework adjusted appropriately.

Outside of the CT context, for example, the legal considerations for targeting operations may be different. The tactical-level processes may also differ; for instance, strike approval may not need to go up to the same level, and the review after the operation may not be conducted in the same way.

Moreover, outside of the CT context, the current public demand for greater legitimacy in targeting operations is minimal.[13] No other types of targeting operations share the notoriety of drone strikes within the CT

---

[13] Note that many of the controversies listed above apply primarily to the CT context.

Introduction | 41

campaign, and it is not even clear that other targeting operations are conducted by the U.S. government with any frequency. As a result, the American public and the international community may afford such actions the benefit of the doubt, with respect to legitimacy. In contrast, drone strike operations continue to be reported, even as the American public's sense of urgency for CT operations fades with each passing year since the 9/11 terrorist attacks. It follows that the decision-making calculus surrounding the framework presented here—weighing the potential benefits of a policy option against the risks—would differ for targeting operations internal and external to the CT context.

Using different methods for targeting, even within the CT context, also involves relevant differences that could significantly change this decision-making calculus. Indeed, other methods of lethal targeting—manned aircraft, teams of Special Operations Forces (SOF), or ship-launched Tomahawk land attack missiles (TLAMs), for example—are not subject to the same controversy or pressure for greater legitimacy that apply to drones. This may be partly a question of perception: drones elicit a visceral negative reaction from many people, where even the word "drone" misleadingly suggests that the aircraft operates autonomously (one reason the term is despised by many in the military). The phenomenon of the public having such a negative perception of drones compared to other means of targeting is also a question of practice. In particular, the U.S. government uses drone strikes in circumstances and with a frequency that other platforms would not permit, as a result of factors such as that drones entail little risk to U.S. forces, may be less expensive than manned platforms, and do not require a ground presence. For all these reasons, the public demand for increased legitimacy for other platforms is limited, hence only a potentially modest public-perception benefit would be gained by implementing a policy option that improved, for instance, the transparency and oversight of operations. At the same time, tactical risks of any policy option would be magnified for targeting operations not done remotely, such as by a manned aircraft strike or a SOF team, because in those cases mission failure could entail the loss of life of a military servicemember or lead to a risky rescue operation.

Hence the framework and analysis presented here may serve as a basis for evaluating policy options for other types of targeting operations, but care should be taken to adapt them as needed.

CHAPTER 8

FRAMEWORK AND POLICY OPTIONS

This chapter introduces an analysis framework and presents the policy options that we will consider in the analysis below.

A FRAMEWORK TO ANALYZE POLICY OPTIONS
The military aspects of the United States' CT campaign can be analyzed along the following lines.

- Four components that become progressively broader in scope:

  - **Tactical military effectiveness (TME)**: the level of success of and considerations related to individual components of an operation;

  - **Operational military effectiveness (OME)**: the level of success of and considerations related to a larger-scale mission or operation, which may entail the actions of multiple units and platforms acting in concert;

  - **Strategic military effectiveness (SME)**: how well military operations contribute to the success of a military campaign, and how well those operations and that campaign further U.S. interests domestically and on the world stage; and

- **Legitimacy**: how U.S. operations, and the military campaign more broadly, are perceived by the American population and the international community.

- How these components influence one another and contribute to the net effectiveness of military actions on the terrorism threat against the United States and its citizens.

The effects of any changes in drone strike policy can therefore be anticipated by considering the impacts on these points. For example, a policy change could influence the TME of drone strikes by affecting the target or strike approval process. Depending on the types of operations in question, a drone strike policy change could influence OME if it affects the quantity and frequency with which drone strikes are carried out, or the coordination between drones and other units and platforms. A drone strike policy change could impact SME by contributing, for instance, to the elimination of a significant terrorist cell or to greater or lesser intelligence sharing between the United States and other countries. Finally, a policy change could impact the perception of the legitimacy of U.S. actions by affecting factors like transparency and accountability.

The discussion that follows focuses primarily on TME and legitimacy. A thorough analysis of the effects of drone strike policy changes on OME and SME is beyond the scope of this work, as OME depends heavily on the details and types of larger-scale operations in which drone strikes are employed, and SME is a very broad concept. These two components would be best addressed by policy makers, analysts, and operators by limiting the focus to those operations and strategic military effects that are of the greatest interest or importance in a given context.

Considerations for evaluating the impacts of a drone strike policy change on TME and perceived legitimacy are presented in the following chapters. These considerations are illustrated through discussions of several specific policy options, which will serve as a theme throughout this work.

POLICY OPTIONS

Numerous policy options for the use of drone strikes in the American CT campaign have been proposed during the last several years. The "military preference" policy, which was issued by President Obama in 2013 (though not yet fully implemented), is one of the policies we consider below. This section also uses as examples a few of the policy proposals put forth that the authors assess as the most viable and most likely to significantly address some of the problematic aspects of strike practices, namely to stand up an oversight body or "drone court" for drone strike operations and to release more detailed information regarding targeting processes and standards and/or targeted individuals. This chapter provides an introduction to these potential policy changes, as well as a preliminary discussion of some of their pros and cons.

MILITARY PREFERENCE

Drone strikes are currently Department of Defense (DOD)-conducted and other government agency (OGA)-conducted.[1] However, OGA strikes have become controversial and have potentially negative legal implications, which are discussed below. In May 2013, President Obama issued classified guidance to implement what was described to the media as a "... preference that the United States military have the lead for the use of force" around the world, regardless of whether operations are in or outside of war zones.[2] This guidance is referred to here as the "military preference" policy. However, in early 2014, Congress at least partially blocked this policy change through a classified annex to the federal budget, so movement in this direction is reportedly happening only very slowly.[3]

---

[1] Rumsfeld Briefing.

[2] "Background Briefing."

[3] Schmitt, "Congress Restricts Drones Program Shift"; Bennett, "McCain Vows New Fight over Control of U.S. Armed Drone Program"; and Mark Mazzetti, "Delays in Effort to Refocus CIA from Drone War," *New York Times*, 5 April 2014, http://www.nytimes.com/2014/04/06/world/delays-in-effort-to-refocus-cia-from-drone-war.html?_r=1.

The military preference policy, as described, would not prohibit OGAs from carrying out drone strikes, but should reduce the number of strikes they carry out (at least, as "lead" organizations). In practice, this would mean that missions previously completed by an OGA using drone strikes will instead be completed using drone strikes with DOD as the lead agency, be completed using means other than a drone strike (e.g., a small SOF team), or not be completed.

This policy guidance has been interpreted less literally by some to be a preference for drone strike operations to be done under Title 10 of the U.S. Code, which would be subtly distinct from the policy just described, and will also be considered in this analysis. In order to explain the distinction, further discussion is required.

Some confusion surrounds the legal regime that governs drone strike (and other related) operations, as evidenced by the common conflation of the military with Title 10 and of intelligence agencies with Title 50 of the U.S. Code. The subtlety that is often missed here centers on "covert actions," which, generally speaking, are actions taken by a government that, at the time the action is carried out, are not intended to be acknowledged by the government.

Title 10 governs the authorities and mechanisms of the armed forces, while Title 50 covers the authorities of the intelligence agencies, intelligence collection, and other such secretive activities. There are important details to note with regard to the interplay between Title 10 and Title 50. Relevant to the discussion here is the fact that covert actions are covered in Title 50—which gives a more nuanced definition than that given above (see Appendix B)—and may only be executed if directed by a presidential finding, but their execution is not restricted to any particular agency. As a result, the military may carry out covert actions—including covert drone strikes—under Title 50. See Appendix A for a more detailed discussion. However, military doctrine provides a definition of covert action that is in some ways broader than the definition given by Title 50. In particular, any military action classified as a "traditional military activity" (TMA) is not covert under U.S. law and therefore falls under Title 10, even though it may be defined by doctrine as covert.

A further complication comes from the fact that the military may act in support of a Title 50 OGA operation. For example, this was the case for the raid that killed Osama bin Laden, which was conducted covertly by military forces, although the raid was technically a Central Intelligence Agency (CIA) operation.[4] Since such an operation is not technically a DOD operation, this is a separate case from that of the military carrying out a Title 50 operation as described above. This practice serves to further blur the association of government departments and agencies with sections of the U.S. Code.

As mentioned above, President Obama's guidance has been interpreted by some to express a preference that drone strike operations be conducted under Title 10 as opposed to Title 50, which would indicate a preference for strikes to be conducted by the military and not to be done covertly (in the Title 50 sense), although it would not limit clandestine operations.[5] While this interpretation may not be supported by the terse explanation issued by the White House, the guidance itself is classified and therefore unavailable for clarification and public debate.[6] Nonetheless, the preference for drone strikes to be conducted under Title 10 is a logical one and worth exploring, regardless of whether it is a part of President Obama's guidance.

Would there be a practical difference between the military preference and this "Title 10 preference"? This depends mainly on whether, in practice, the military engages in covert actions, as defined in Title 50 as the

---

[4] "CIA Chief Panetta: Obama Made 'Gutsy' Decision on Bin Laden Raid," PBS Newshour (video and transcript), 3 May 2011, http://www.pbs.org/newshour/bb/terrorism-jan-june11/panetta_05-03.html.

[5] A clandestine operation is defined doctrinally as "an operation sponsored or conducted . . . in such a way as to assure secrecy or concealment." Military doctrine goes on to clarify: "A clandestine operation differs from a covert operation in that emphasis is placed on concealment of the operation rather than on concealment of the identity of the sponsor." See *Department of Defense Dictionary of Military and Associated Terms* Joint Publication 1-02 and Joint Publication 3-05.1 *Joint Special Operations: Task Force Operations*. U.S. Joint Chiefs of Staff, Appendix B discusses issues related to conducting drone strikes in a clandestine or covert manner.

[6] See "Background Briefing" for the explanation from the White House.

lead, as opposed to the supporting, agency. Legally it may, as mentioned above, and the analysis below considers this case.

Throughout the following, a preference for drone strike operations to be conducted under Title 10 is considered as a potential option for the military preference policy. It is discussed specifically in the cases when it has different or additional implications than does the military preference in the absence of the additional Title 10 preference.

If the military does not carry out covert actions, this would simplify some of the discussion that follows. Even so, note that the additional Title 10 preference would make for a more transparent policy. As discussed below, there are advantages to transparency, but there might also be military and strategic costs.

Speaking of transparency, even this relatively brief discussion highlights some of the issues around President Obama's guidance being classified. The public release of what specifically it entails, assuming it is or will be implemented, would significantly increase the transparency of U.S. drone strike operations. Another policy option discussed below entails the release of additional information about targeting processes and standards, and could include the release of the guidance or at least more information describing it. The pros and cons of that general policy option will be considered throughout Part II, alongside those of the military preference policy.

A "DRONE COURT"

The push for increased accountability and oversight has resulted in calls for a "drone court" to oversee drone strikes. Although drones have caught the public's attention, it may be more logical to expand this notion to that of a court that oversees remote targeting operations, rather than the court's purview being dependent on the weapons platform being a drone. Moreover, such a mechanism for oversight could potentially exist outside of the judicial branch so, even though the word "court" is used throughout the following, this option should be understood to be an oversight body rather than a court necessarily. Thus the term "drone court" will be used for our purposes as a short-hand reference to a body that oversees remote targeting operations in which people are killed, including those operations carried out by drones.[7]

A drone court could take a couple of different forms. One form that is widely discussed is the notion of a "FISA-like drone court," referring to the court established by the Foreign Intelligence Surveillance Act (FISA). The federal judges presiding over the FISA court approve warrants permitting U.S. agencies (in particular, the National Security Agency [NSA] and the Federal Bureau of Investigation [FBI]) to collect evidence—typically electronic communications—pertaining to foreign intelligence and terrorism. The analog in the drone world would be a court in which judges authorize drone strikes or drone strike targets based on whether they are legally sound and/or in line with U.S. policy, except in cases for which time does not permit, wherein a post-strike review would be conducted. As with the FISA court, the proceedings would be classified and held out of the public view. The court could be limited to overseeing only operations located in areas where American ground troops have no presence or have jurisdiction over all operations.

This option has received high-profile attention. Members of the U.S. Senate Select Committee on Intelligence, including Chairman Dianne Feinstein and Senator Angus S. King, as well as Senate Judiciary Committee Chairman Patrick J. Leahy and Ranking Member Charles E. Grassley have expressed interest in it.[8] Former Secretary of Defense and CIA Director

---

[7] A drone court could alternatively have oversight over targeting operations in general. This may include operations in support of more widely varied missions than if the purview of the court was strictly over remote targeting operations.

[8] Carlo Muñoz, "Sens. Feinstein, Leahy Push for Court Oversight of Armed Drone Strikes," *The Hill*, 10 February 2013, http://thehill.com/policy/defense/282033-feinstein-leahy-push-for-court-oversight-of-armed-drone-strikes-; John O. Brennan, "Open Hearing: Nomination of John O. Brennan to be the Director of the Central Intelligence Agency," 7 February 2013 (transcript and video), United States Senate, Select Committee on Intelligence, Washington, DC, http://www.intelligence.senate.gov/hearings/open-hearing-nomination-john-o-brennan-be-director-central-intelligence-agency, hereafter Brennan, "Open Hearing"; and Scott Shane, "Debating a Court to Vet Drone Strikes," *New York Times*, 8 February 2013, http://www.nytimes.com/2013/02/09/world/a-court-to-vet-kill-lists.html.

Robert M. Gates expressed support for this option, and President Obama stated his intention to review its merits.[9] However, a FISA-like court raises a number of significant legal and practical questions, such as whether it would be constitutional or an unprecedented intrusion into battlefield decision making and the president's prerogatives as commander in chief.

Another alternative would be to establish an "Israeli-style drone court" to review drone strikes after they have occurred. Such a system has been in place in Israel for several years, in which, as stipulated by the Israeli Supreme Court, after a targeted killing strike takes place, an independent body carries out "a thorough investigation regarding the precision of the identification of the target and the circumstances of the attack upon him . . ."[10] These proceedings are not accessible by the public, but add some level of oversight and accountability to the process.[11] Currently, U.S. government organizations that carry out drone strikes conduct internal investigations according to their own criteria (or sometimes at the direction of the relevant congressional committees).

There are many different potential ways to implement an Israeli-style court. It could evaluate strikes based on their adherence to legal or policy standards. It could release its findings to the public or not. It could investigate all targeted killings, or only those alleged to have been improper or that resulted in unintended civilian casualties. As with the FISA-like court, its jurisdiction could be restricted to operations conducted outside of hot

---

[9] "Drone Court Proposed to Review Targeted Killings of Americans," Arutz Sheva 7, 10 February 2013, http://www.israelnationalnews.com/News/News.aspx/165081#.VcTUanj9pi0; and Obama, "Remarks by the President at the National Defense University."

[10] Public Committee against Torture in Israel et al. v. Government of Israel et al. (High Court of Justice, 2005), http://elyon1.court.gov.il/Files_ENG/02/690/007/a34/02007690.a34.htm. This Israeli Supreme Court decision was a landmark one from 2006 in which the court held that it was legal for the military to execute targeted killings against members of designated terrorist organizations.

[11] For a description of some of the general mechanisms for oversight and investigation used in Israel, see *Israel's Mechanisms for Examining and Investigating Complaints and Claims of Violations of the Laws of Armed Conflict According to International Law: Second Report-The Turkel Commission* (Israel: Public Commission to Examine the Maritime Incident of 31 May 2010, 2013), http://www.turkel-committee.gov.il/files/newDoc3/The%20Turkel%20Report%20for%20website.pdf.

battlefields, or not. Under this system, the government could have the opportunity (or be compelled) to release—to Congress or the pubic—its findings that the target was guilty of wrongdoing. Finally, this system could also serve as an instrument to award monetary reparations in response to civilian casualties and other collateral damage. As with a FISA-like drone court, the Israeli-style model also raises significant legal and practical questions.

The most obvious way to run a FISA-like court or an Israeli-style court would be to populate it with federal judges, keeping it within the judicial branch, although it would raise some questions about legality and whether the judges would necessarily have the appropriate national security expertise. Alternatively, either entity could be run as an independent oversight board within the executive branch, which President Obama referred to in a May 2013 speech, without providing further details.[12] Another option would be to implement the drone court as a "national security court"— a hybrid between a federal court and a military commission—existing outside of the federal court system and modeled after such a proposal for dealing with detainees.[13] The court would be overseen by civilian appointees with expertise in national security issues.[14] Both of these options might have significant potential, but present a significant number of legal and practical issues that are outside the scope of this work.

RELEASING FURTHER DETAILS ABOUT TARGETING

The Obama administration has put forth basic information about its drone strike targeting practices in various speeches and other communiqués, but the demand for greater transparency continues to grow. The Department of Justice (DOJ) was recently compelled by a Freedom of Information Act

---

[12] Obama, "Remarks by the President at the National Defense University."

[13] Marc Ambinder, "The National Security Court System: An Interview with Glenn Sulmasy," *Atlantic*, 5 August 2009, http://www.theatlantic.com/politics/archive/2009/08/the-national-security-court-system-an-interview-with-glenn-sulmasy/22708/.

[14] Glenn M. Sulmasy, "The Legal Landscape after Hamdan: The Creation of Homeland Security Courts," *New England Journal of International and Comparative Law* 13, no. 1 (Fall 2006).

(FOIA) lawsuit to release portions of a classified memorandum that put forth its legal rationale for targeting U.S. citizen Anwar al-Awlaki, while members of the public and even of the Senate Select Committee on Intelligence pushed for the release of additional DOJ legal memos on drone strike practices.[15]

More widely, the UN, Congress, advocates, and pundits have called for the release of further details on U.S. drone strike activities and practices. In the 2014 Intelligence Authorization Act, Congress proposed a provision that would have required the president to report the total number of combatants and noncombatant civilians killed or injured by drone strikes in the past year, although it later stripped the provision from the bill.[16] With respect to U.S. targeting processes and standards, Senators Ronald L. Wyden, Mark E. Udall, and Martin T. Heinrich of the Senate Select Committee on Intelligence noted in a public letter to Attorney General Holder, "The United States' playbook for combating terrorism will sometimes include sections that are secret, but the rulebook that the United States follows should always be available to the American public."[17]

While the primary focus of these members of Congress may be on the release of information to the American people, greater transparency to the local population in areas of drone strikes may also be desirable. For

---

[15] Anwar al-Awlaki was an American citizen and a member of al-Qaeda in the Arabian Peninsula who was actively involved in terrorist activities. He was targeted and killed by a drone strike in Yemen in 2011. See Brennan, "Open Hearing." For more on the DOJ release, see Charlie Savage, "Court Releases Large Parts of Memo Approving Killing of American in Yemen," *New York Times*, 23 June 2014, http://www.nytimes.com/2014/06/24/us/justice-department-found-it-lawful-to-target-anwar-al-awlaki.html. A request came from senators for the release of additional DOJ memos, see "Letter from Senators Ron Wyden, Mark Udall, and Martin Heinrich to the Honorable Eric Holder," 26 November 2013, http://www.wyden.senate.gov/download/?id=C48CD5E5-EF15-4A44-A1BF-2274E5B1929A&download=1.

[16] "Drone Strikes: James Clapper's Letter to Senate Intelligence Committee," *Guardian*, 29 April 2014, http://www.theguardian.com/world/interactive/2014/apr/29/cia-us-national-security; and Spencer Ackerman, "U.S. Senators Remove Requirement for Disclosure over Drone Strike Victims," *Guardian*, 28 April 2014, http://www.theguardian.com/world/2014/apr/28/drone-civilian-casualties-senate-bill-feinstein-clapper.

[17] "Letter from Senators Ron Wyden, Mark Udall, and Martin Heinrich to the Honorable Eric Holder."

example, "In a place like Yemen," one article notes, "although the American drone program is universally hated, many Yemenis will admit they would support targeted assassinations if there is clear intelligence that an individual is a senior operative within AQAP [al-Qaeda in the Arabian Peninsula] and plotting a specific and imminent act of terror against Americans."[18]

As one legal scholar asserted, "The government needs a way to credibly convey to the public that its decisions about who is being targeted . . . are sound."[19] While certain specifics of U.S. targeting practice and policy should remain protected to preserve the effectiveness of the intelligence collection methods and the strikes themselves, some materials might be able to be released in a way that does not prohibitively harm U.S. national security interests. Two separate (and independent) options are considered here: (1) releasing more information about the targeting process itself and targeting standards; and (2) releasing post-strike details about specific targets. Such information should be released only after any potential harm to the future effectiveness of U.S. practices has been fully weighed.

Specific points of clarification for targeting processes and standards could include:

- which agencies and how many people take part in the target approval process;[20]

- what the intelligence review processes are like, in general;[21]

---

[18] Danya Greenfield, "The Case Against Drone Strikes on People Who Only 'Act' Like Terrorists," *Atlantic*, 19 August 2013, http://www.theatlantic.com/international/archive/2013/08/the-case-against-drone-strikes-on-people-who-only-act-like-terrorists/278744/.

[19] "Drone Court Proposed to Review Targeted Killings of Americans."

[20] Jack Goldsmith, "Fire When Ready," *Foreign Policy*, 20 March 2012, http://foreignpolicy.com/2012/03/20/fire-when-ready/.

[21] Ibid.

- further explanation of the requirement that a target represent an "imminent" threat;[22]

- provide more detail on what constitutes the "infeasibility" of capture for a target;[23]

- what the means are for deciding, for targeting purposes, when members of al-Qaeda and its associated forces are performing continuous combat functions, and when civilians are directly participating in hostilities;

- the level of evidence the president needs to determine that a given American may be targeted by military action;[24]

- the level of evidence that is needed to target unknown individuals in a "signature" strike;

- provide further guidance on what types of military activities are considered TMAs (i.e., are conducted under Title 10, rather than Title 50);[25]

- what the congressional review entails, including distinctions between the Intelligence and Armed Services Committees' processes;[26] and

- what President Obama's May 2013 guidance entails and, to the extent that it is being implemented, what changes it has introduced into U.S. CT practices.

The other option is to release some of the target's terrorist affiliations, activities, and plans after a strike is completed, when it is feasible to do so without revealing sources and methods in a way that would significantly

---

[22] "Letter from Senators Ron Wyden, Mark Udall, and Martin Heinrich to the Honorable Eric Holder."
[23] Ibid.
[24] Ibid.
[25] See Appendix A for further discussion of these points.
[26] For more on the congressional review. see Goldsmith, "Fire When Ready."

hinder future operations and intelligence collection. The information released could include:

- details of the target's ties to al-Qaeda or an associated force;
- explanation of the military necessity of the strike and why the threat posed was imminent; and
- explanation of why capture was not feasible.[27]

Currently, this type of information is only being released in extremely rare cases. For instance, some of these details were addressed in the case of al-Awlaki.[28] Note that an Israeli-style drone court would be one potential way to implement this option.

SUMMARY OF POLICY OPTIONS

The five policy options described above are revisited in each of the remaining chapters of Part II. The options are summarized as follows:

- **The military preference:** a preference that drone strikes be carried out by the military. This option may include the additional preference that drone strikes be carried out under Title 10 (i.e., that drone strikes not be conducted covertly as defined by Title 50).

- **A FISA-like drone court:** establishing a process by which drone strikes are authorized by an oversight body (with exceptions for cases in which time is too short).

- **An Israeli-style drone court:** establishing a process that reviews drone strikes after the fact.

- Releasing further details about targeting processes and standards.

- Releasing further details about targeted individuals.

---

[27] Ibid.
[28] *Joint Special Operations: Task Force Operations.*

CHAPTER 9

MILITARY EFFECTIVENESS

The remainder of Part II describes considerations for anticipating the effects of policy options for drone strikes on military effectiveness (with a focus on tactical military effectiveness or TME), perceived legitimacy, and how those factors contribute to the net effectiveness of combatting terrorism. The policy options described in the previous chapter serve as examples throughout these discussions.

Effects on TME, OME, and SME are crucial factors to consider when evaluating any drone strike policy option. Presented below is one way to think through the effects of a policy option on TME, as well as considerations related to adversary reactions, which are relevant to all three aspects of military effectiveness.

TME

One way to anticipate the effects on TME due to a drone strike policy option is to break a single generic strike down into tactical-level steps and consider the effects on each step individually. Consider the following seven tactical steps for drone strikes:

1. **Targeting:** intelligence products are used to designate a target based on established criteria.

2. **Approval of target:** the appropriate authority approves the target, which can happen in conjunction with "approval of strike" step.

3. **Plan strike:** the strike is planned, coordinated, and deconflicted by the forces that will conduct it, and resources are apportioned for it. The plan requires a drone based within reach of the strike area, either on land or afloat, and the use of airspace between the base and the strike area.

4. **Approval of strike:** the appropriate authority approves the strike, which can happen months, days, or hours before the strike is executed.

5. **Conduct strike:** the strike is carried out in accordance with the plan, as well as the standard procedures set by the organization(s) in command and control of the operation.

6. **Immediate review:** a battle damage assessment (BDA) may be conducted, either by surveillance assets like a drone or by individuals on the ground. The strike may be debriefed by operators, and lessons learned may be extracted.

7. **Longer-term review and reaction:** the outcome of the strike may be reported to oversight authorities, such as Congress. If there are significant concerns about the strike or its outcomes, an internal or external investigation may be conducted and could result in disciplinary actions, acknowledgement of collateral damage, or changes to training or operations. Any lessons learned may also be absorbed and reflected in adaptations made to training or operations.

Note that the immediate and longer-term review steps may entail any subset of the actions listed or none at all.

The specifics of how each step is carried out will depend on the organization(s) carrying out the operation and the details of the operation itself. These steps will also vary depending on whether the strike is planned in advance or arises out of a dynamic situation, such as an opportunity for a spontaneous strike that would further the objectives of the military campaign, but for which the target or strike itself had not previously been approved (e.g., in the case of a signature strike). DOD doctrine refers to these as "unanticipated" or "unplanned" targets of opportunity, respectively.[1] In those cases, the target and strike planning and approval processes might be based on general guidelines, with a more robust review after the strike takes place.

DOD doctrine supports the division of the targeting process into generic steps. In particular, these steps reflect those from the joint targeting cycle, which are the following:[2]

1. **Endstate and commander's objectives**: forces ensure that the commander's intent, the conditions that characterize the military objectives have been met, and corresponding metrics are all developed and thoroughly understood by forces.[3]

2. **Target development and prioritization**: the adversary's systems of interest are analyzed to determine potential targets and those targets are developed, prioritized, nominated, and approved.

3. **Capabilities assessment**: forces evaluate their own capabilities in the context of perhaps numerous anticipated target requirements and determine and analyze options for prosecuting those targets.

4. **Commander's decision and force assignment**: options from the previous step are further analyzed with respect to the available forces, systems, and necessary support; the commander approves the target list; and forces are tasked.

5. **Mission planning and force execution:** operations are planned in detail, adjusted as necessary in reaction to any changing conditions, executed, and an initial BDA is completed.

6. **Assessment:** the commander evaluates the effect of the actions from the five previous steps on achieving the necessary objectives, does further BDA, and recommends follow-on actions.

---

[1] *Joint Targeting*, Joint Publication 3-60 (Washington, DC: Joint Chiefs of Staff, 2013).

[2] Ibid.

[3] This step is not a part of the process described above.

Several of the seven tactical steps for drone strikes discussed earlier have been expounded upon by government officials. For example, in a 2012 speech, John Brennan, then-chief counterterrorism advisor to President Obama and the current director of the CIA, noted that the target approval step entails an evaluation of whether the potential target is a "significant threat to U.S. interests" and goes up to the "most senior officials in our government."[4] This statement was supported by interviews with several DOD officials, who specified that a group of senior officials from such agencies as the Departments of State and Justice review the relevant intelligence and approve or reject DOD targets. The officials also noted that DOD's target approval process entails interagency legal vetting.

Minimizing civilian drone strike casualties has always been a priority of the United States, and President Obama recently specified that the current policy is that strikes are carried out only if there is "near certainty that no civilians will be killed or injured."[5] Furthermore, a planned strike receives approval only if U.S. forces have a "high degree of confidence" in the identity of the target.[6] Immediate review includes BDA and, if collateral damage occurred, then longer-term review and reaction may include analyzing the strike process and making changes based on those findings.[7]

Effects of the Policy Options

The military preference, FISA-like drone court, and Israeli-style drone court policies each have an impact on the tactical steps of a drone strike. These implications should be evaluated fully by those currently involved

---

[4] John O. Brennan, "The Ethics and Efficacy of the President's Counterterrorism Strategy" (remarks, Washington, DC, 30 April 2012), http://www.fas.org/irp/news/2012/04/brennan 043012.html.

[5] Regarding minimizing casualties, see ibid.; and Senior Defense Official, "Background Briefing on Targeting," DOD News, 5 March 2003, http://www.defense.gov/Transcripts/Transcript.aspx?TranscriptID=2007. Quotation from Obama, "Remarks by the President at the National Defense University."

[6] Brennan, "The Ethics and Efficacy of the President's Counterterrorism Strategy."

[7] Ibid.

in drone strike operations. Some initial considerations along these lines follow.

For the military preference policy, note that how each of the seven tactical steps is carried out might differ depending on the agency leading the strike. Take, for instance, "longer-term review and reaction": oversight for many DOD drone strikes is with the House and Senate Armed Services Committees, while a drone strike carried out by an OGA receives oversight by the House and Senate Intelligence Committees.[8] Also, each organization that carries out drone strikes follows its own procedures, so the mechanics for strike planning and execution will vary depending on whether strike planning and execution are conducted by the DOD or an OGA. DOD and OGA operators and other relevant personnel should be able to predict the effects that the military preference policy would have on each of these steps.

The necessary regional access to carry out drone strikes also might vary depending on the organization conducting the strikes. For example, OGAs apparently have more flexibility to operate in Yemen than the U.S. military does. Indeed, as a result of a number of drone strikes that mistakenly killed innocent people, Yemen banned the U.S. military from conducting drone strikes in the country, though OGA strikes reportedly continued.[9] This example shows that the military preference policy might hinder the U.S.'s ability to launch strikes depending on location.

Moreover, the drone strike process looks different depending on whether a strike is covert, so there would be further ramifications if the military preference policy included a preference that drone strikes be conducted under Title 10 (i.e., not be conducted covertly, in the Title 50 sense). Indeed, a "Title 50 covert" strike requires a presidential finding for approval, whereas other strikes need not rise to this level of authorization. Oversight of a "Title 50 covert" strike must include the intelligence committees, and any investigations or acknowledgement of collateral damage could, by

---

[8] Whether there is a substantive difference between the oversight of the different congressional committees is thus an important question, which publicly available information may not answer.

[9] Mazzetti, "Delays in Effort to Refocus CIA from Drone War."

definition, only be internal to the government in the case of a covert strike. Again, DOD and OGA operators and other relevant personnel should be able to estimate the effects of these factors.

A FISA-like drone court would affect the target approval process, with the court itself adding another layer of oversight and time via its deliberations to ensure that the legal basis for the proposed action is sound. While those factors could hinder fast and streamlined execution of strikes, they could potentially serve as an additional check against collateral damage and other unintended consequences. An Israeli-style drone court would add additional accountability and perhaps a mechanism for addressing collateral damage in the longer-term review and reaction processes. More accountability could potentially have a chilling effect on operations as well, with the government less inclined to carry out drone strikes in case they would result in sanctions or additional controversy.

Finally, all of the policy options considered here could have an effect on the cooperation of U.S. allies and other nations with U.S. drone strike operations, including, for example, drone basing and airspace usage, as mentioned in the planning step. These aspects of drone strike operations are discussed more holistically below.

THE ENEMY'S VOTE

"The enemy gets a vote" in the outcome of any military action, as the saying goes. Consider a single drone strike, which entails the process described above. The actions of the target and his or her network may make the steps of that process easier or harder to accomplish successfully. The target's ability to affect the strike process is based, in part, on their understanding of U.S. methods of intelligence collection, operations, and procedures. This topic affects TME, OME, and SME.

For the targeting stage, collecting intelligence to identify the target and his or her terrorist activities and affiliations is crucial and, in order to plan the strike, establishing intelligence on the local area and the target's pattern of life is necessary. All of this depends on U.S. forces' intelligence-gathering capabilities being effective in the target's local area, and the target (and their network) not being able to or aware enough to avoid these efforts. Intelligence-sharing agreements with U.S. allies can also provide an essential complement to these capabilities.

The strike must be conducted in accordance with the executing organization's standards and procedures. The more the target knows about these, the more they would be able to ensure those standards are not met, preventing the strike from taking place. For example, President Obama has said in recent years of current U.S. policy that, ". . . before any strike is taken, there must be near-certainty that no civilians will be killed or injured . . ."[10] Hence, if a terrorist can ensure that they are clearly in the presence of civilians, they may be able to avoid a strike. In that vein, the more terrorists know about the target approval and accountability processes for strikes, and the more robust those processes are, the more the terrorists may be able to deduce ways of making U.S. forces less inclined to act, either because a strike is not approved or for fear of repercussions under the accountability processes. Conversely, less transparency about targeting practices could provide strategic ambiguity around U.S. actions, which could be exploited to deter potential terrorists.

Effects of Policy Options
The obvious danger of policy options that consist of releasing more information on U.S. targeting processes and standards or potential targets is that terrorists might be able to use this information to be better able to thwart intelligence-gathering and strike operations. This issue presents a significant risk to the military effectiveness of drone strikes that should be taken into account when analyzing options.

As for the military preference policy, one of the main arguments for the policy is that DOD carries out operations under a relatively well-understood collection of standards and procedures. Former Director of National Intelligence (DNI) Admiral Dennis C. Blair noted that, "Within the armed forces we have a set of procedures that are open, known for how

---

[10] Obama, "Remarks by the President at the National Defense University."

you make decisions about when to use deadly force or not, levels of approval, degrees of proof and so on and they are things that can be and should be out."[11] The extent to which the transparency given by drone operations comporting to those standards and enabling terrorists to avoid drone strikes is similarly a factor that should be taken into account.

Finally, the additional approval authority of a FISA-like drone court, or the additional accountability measure of an Israeli-style drone court, might also prevent the United States from acting in certain instances, which could potentially be manipulated by terrorists. The extent to which this could happen will depend on the standards of the court(s) as well as the transparency of those standards.

SUMMARY OF MILITARY EFFECTS ISSUES

The discussion above provides an avenue for evaluating the potential effects of a policy change for drone strikes on TME by considering the various processes of a single drone strike. The generic drone strike steps highlight specific areas to consider within TME when evaluating drone strike policy options. To make a concrete evaluation, input from current operators and the details of how the policy options would be implemented would be required.

This chapter also suggests a significant area in which a policy change could indirectly impact military effectiveness at the tactical, operational, and strategic levels—that of the terrorists' abilities to avoid being targeted. Indeed, the more individuals know about drone strike practices and targeting guidelines, the more they may be able to avoid meeting the criteria for a strike. Since all of the policy options presented involve some degree of greater transparency for drone strike operations, they all risk giving terrorists this type of advantage, particularly the options for releasing more information about targeting processes and standards and targeted individuals.

---

[11] Scott Horton, "Blair Addresses the CIA, Drones, and Pakistan," *Browsings* (blog), *Harper's*, 1 December 2011, http://harpers.org/archive/2011/12/hbc-90008329.

CHAPTER 10

LEGITIMACY

U.S. actions abroad are considered legitimate to a given audience to the extent that they are in line with that group's values and perceived norms. For instance, the government will garner legitimacy in the eyes of the West if U.S. actions are consistent with Western values (e.g., transparency and advancing personal and economic freedoms) and international law, which perhaps forms the perceived set of norms in the West. To the international community writ large, common values might be more basic, including such principles as fairness, accountability, and ethical standing, and perhaps only portions of international law form the common perceived norms. The values of the American public combine Western values with an American worldview. For example, after years as an active global hegemon, an American worldview might be more interventionist than that of the West as a whole.

This chapter discusses the international legal context in which drone strikes are carried out. It subsequently provides the details of the "legitimacy" portion of our framework by specifying numerous legitimacy issues, particularly how these issues are addressed by the five policy options.

INTERNATIONAL LEGAL CONTEXT

The U.S. CT campaign against al-Qaeda and its associated forces is carried out within the context of international law. Whether drone strikes are seen to be consistent with international law forms a major component of their perceived legitimacy throughout the world.[1] This chapter provides background on the existing international legal framework and how the U.S. government describes its drone strike practices as fitting into that framework. This chapter also highlights some controversies and issues that have emerged around the U.S. CT campaign and drone strikes more specifically, as well as the role that the policy options considered here could play in addressing some of those issues. Note that the issues raised within the more in-depth legal discussion here are briefly summarized in the next section.

The government has stated unequivocally that its use of drone strikes is consistent with international law.[2] In particular, the government has asserted that its use of drone strikes in CT operations qualifies as a legal use of force because it is a self-defense response to the 9/11 terrorist attacks and the continuing terrorist threat from al-Qaeda and its associated groups.[3] Indeed, on 12 September 2001, the United Nations Security Council passed

---

[1] This is true even though international law has significant areas of controversy and ambiguity. For example, the Geneva Conventions—consisting of four Conventions and three Additional Protocols—form the basis for important aspects of international law. Even so, many countries have not ratified one or more of the Additional Protocols, including the United States, which has not ratified the first and second of the Protocols. However, some articles from the Protocols have been deemed customary law, and thereby apply to all nations.

[2] See Obama, "Remarks by the President at the National Defense University"; Brennan, "The Ethics and Efficacy of the President's Counterterrorism Strategy"; and Eric H. Holder Jr., "Attorney General Eric Holder Speaks at Northwestern University School of Law" (speech, Northwestern University School of Law, 5 March 2012), http://www.justice.gov/iso/opa/ag/speeches/2012/ag-speech-1203051.html. While the U.S. government also has stated unequivocally that its drone strike practices are consistent with domestic law, this work focuses primarily on international law.

[3] See, for example, Brennan, "The Ethics and Efficacy of the President's Counterterrorism Strategy"; Holder, "Attorney General Eric Holder Speaks at Northwestern University School of Law"; and Harold Hongju Koh, "The Obama Administration and International Law" (speech, Annual Meeting of the American Society of International Law, Washington, DC, 25 March 2010), http://www.state.gov/s/l/releases/remarks/139119.htm. Note that the UN Charter permits the use of force against a threat within another nation's borders under three circumstances: for self-defense (including collective self-defense), with the consent of the host nation, or in accordance with a Security Council resolution. In the case of self-defense, the host nation must be unable or unwilling to act against the threat.

a resolution condemning the 9/11 attacks and noting a nations' right to self-defense.[4]

The U.S. government contends that it is involved in an armed conflict with al-Qaeda and its associated forces—specifically, what some term a "transnational" Noninternational Armed Conflict (NIAC).[5] The government's assertion is not without controversy, however. For instance, the International Committee of the Red Cross (ICRC) takes issue with the United States classifying all its actions against al-Qaeda and its associated forces in this way. ICRC's position is that each "situation of violence" should be considered separately and classified as an International Armed Conflict (IAC), an NIAC, or not an armed conflict on a case-by-case basis, and that situa-

---

[4] See UN Security Council Resolution (UNSCR) 1368, UN Doc. S/RES/7143, 12 September 2001, http://www.un.org/News/Press/docs/2001/SC7143.doc.htm; and UNSCR 1373, UN Doc. S/RES/7158, 28 September 2001, http://www.un.org/News/Press/docs/2001/sc7158.doc.htm. On the domestic front, the U.S. Congress passed the Authorization for the Use of Military Force (AUMF) on 14 September 2001, which permitted U.S. military action against the individuals, organizations, and nations that perpetrated or aided in the 9/11 attacks without geographic or temporal limitations.

[5] Obama, "Remarks by the President at the National Defense University"; Brennan, "The Ethics and Efficacy of the President's Counterterrorism Strategy"; Koh, "The Obama Administration and International Law"; and Emmerson, *Promotion and Protection of Human Rights*. International law has differing standards for the use of force within and outside of armed conflicts, and also distinguishes between IACs and NIACs. A conflict is an IAC if it occurs between nations or if a population is defending its right of self-determination against "colonial domination and alien occupation and . . . racist regimes." See the *Protocol Additional to the Geneva Conventions of 12 August 1949*, opened for signature 12 December 1977, UN Doc. A/32/144. Outside of these "national liberation" cases, a conflict is a NIAC if one or more of the main parties to the conflict is an organized nonstate group, meaning that the group has ". . . certain command structure and the capacity to sustain military operations," and the conflict is protracted and is of a level of intensity above that of "internal disturbances and tensions, such as riots." See *How is the Term "Armed Conflict" Defined in International Humanitarian Law?*, ICRC Opinion Paper (Geneva, Switzerland: ICRC, 2008), http://www.icrc.org/eng/assets/files/other/opinion-paper-armed-conflict.pdf; and the *Protocol Additional to the Geneva Conventions of 12 August 1949*, and *relating to the Protection of Victims of Non-International Armed Conflicts (Protocol II)*, opened for signature 12 December 1977, UN Doc. A/32/133. (The notion of a "transnational" NIAC is relatively new.) A conflict is not considered an armed conflict if it does not fall into these two categories: if violent events are sporadic or rise only to the level of intensity of a disturbance or if parties to the conflict are not sufficiently organized.

tions of violence in the U.S. CT campaign have fallen under each of these categories.[6]

International Humanitarian Law (IHL) governs the use of force in armed conflicts. IHL stipulates, for example, that individuals can be targeted in an armed conflict if they are members of the armed forces of a party to the conflict or additionally, in the case of a NIAC, if they are members of an organized armed group that is a party to the conflict and have a continuous function to directly participate in hostilities.[7] In addition, civilians may be targeted if they are directly participating in hostilities, subject to meeting specific criteria regarding the nature of their actions.[8] Any such targeting or use of force is further subject to the principles of military necessity, distinction, and proportionality under IHL.[9]

The Obama administration has argued that its targeting practices conform to these principles, and has outlined further aspects of U.S. targeting to make the case that it is in line with IHL and to some extent with International Human Rights Law (IHRL), which includes a more stringent

---

[6] *International Humanitarian Law and the Challenges of Contemporary Armed Conflicts* (Geneva, Switzerland: 31st International Conference of the Red Cross and Red Crescent, 28 November –1 December 2011), http://www.icrc.org/eng/resources/documents/report/31-international-conference-ihl-challenges-report-2011-10-31.htm.

[7] Nathalie Weizmann, "Remotely Piloted Aircraft and International Law," in *Hitting the Target? How New Capabilities Are Shaping International Intervention*, ed. Michael Aaronson and Adriand Johnson (London: Royal United Services Institute for Defence and Security Studies, 2013), http://www.icrc.org/eng/resources/documents/article/other/remotely-piloted-aircraft-ihl-weizmann.htm; and Nils Melzer, *Interpretive Guidance on the Notion of Direct Participation in Hostilities under International Humanitarian Law* (Geneva: ICRC, 2009), hereafter *Interpretive Guidance*, http://www.icrc.org/eng/resources/documents/publication/p0990.htm.

[8] Weizmann, "Remotely Piloted Aircraft and International Law"; and Melzer, *Interpretive Guidance*. The topic of direct participation in hostilities is nuanced and an area of debate beyond the scope of this work.

[9] Necessity requires that force only be used lawfully and to the level needed to achieve the military objective. The principle of distinction speaks to discerning between legitimate military targets and protected entities (e.g., civilians who are not taking part in hostilities or civilian infrastructure); attacks must attempt to limit damage to legitimate targets when possible, and armed forces should attempt to distinguish and separate themselves and their fixtures from the civilian population. Proportionality requires that collateral damage be limited to a level "proportional" to the military objective sought. In addition, some cite the principle of humanity, which requires the use of weapons that do not inflict unnecessary suffering.

set of standards for the use of force that applies even outside of armed conflicts.[10] In particular, IHRL allows individuals to be targeted with force only if they represent an imminent and substantial threat to life, and arrest is not reasonably possible.

The Obama administration provided some clarification as to who qualifies as a targetable member of al-Qaeda and its associated groups. In a 2012 speech, then-DOD General Council Jeh C. Johnson stated:

> An "associated force," as we interpret the phrase, has two characteristics to it: (1) an organized, armed group that has entered the fight alongside al Qaeda, and (2) is a co-belligerent with al Qaeda in hostilities against the United States or its coalition partners. In other words, the group must not only be aligned with al Qaeda. It must have also entered the fight against the United States or its coalition partners.[11]

Harold Koh, then legal advisor to the Department of State, stated that U.S. forces' criteria for determining individual membership in al-Qaeda or an associated group "includes, but is not limited to . . . relevant evidence of formal or functional membership, which may include an oath of loyalty, training with al-Qaeda, or taking positions with enemy forces."[12]

Harkening back to IHRL standards, lethal force is used only if the target "poses a continuing, imminent threat to U.S. persons," and only then when capture is not feasible and there are "no other reasonable alter-

---

[10] For the Obama administration's argument, see Brennan, "The Ethics and Efficacy of the President's Counterterrorism Strategy." The listed aspects of IHRL can be considered regulation on law enforcement.

[11] Jeh Charles Johnson, "National Security Law, Lawyers, and Lawyering in the Obama Administration" (dean's lecture, Yale Law School, 22 February 2012), http://www.lawfareblog.com/2012/02/jeh-johnson-speech-at-yale-law-school/.

[12] Koh, "The Obama Administration and International Law."

natives" to address the threat posed.[13] Attorney General Holder spoke to the feasibility of capturing a terrorist suspect when that suspect is a U.S. citizen, although that criteria could potentially apply to suspects of any nationality:

> Whether the capture of a U.S. citizen terrorist is feasible is a fact-specific, and potentially time-sensitive, question. It may depend on, among other things, whether capture can be accomplished in the window of time available to prevent an attack and without undue risk to civilians or to U.S. personnel. . . . In that case, our government has the clear authority to defend the United States with lethal force.[14]

When he was the chief counterterrorism advisor in the Obama administration, John Brennan described aspects of targeting processes and standards.[15] A potential target and suspected al-Qaeda member is vetted to determine lawfulness, and in that case presented to "the very most senior officials in our government" for evaluation. Interviews conducted with several DOD officials confirmed that the target nomination meetings are lively interagency interactions at an extremely high level. Brennan noted that one criterion considered during the target approval process is whether the individual poses "a significant threat to U.S. interests," for instance, acts as operational leader, makes preparations to attack U.S. interests, or "possesses unique operational skills that are being leveraged to plan an attack."[16]

---

[13] The White House, Office of the Press Secretary, "Fact Sheet: U.S. Policy Standards and Procedures for the Use of Force in Counterterrorism Operations Outside the United States and Areas of Active Hostilities," 23 May 2013, hereafter "Fact Sheet," https://www.whitehouse.gov/the-press-office/2013/05/23/fact-sheet-us-policy-standards-and-procedures-use-force-counter-terrorism.

[14] Holder, "Attorney General Eric Holder Speaks at Northwestern University School of Law."

[15] Brennan, "The Ethics and Efficacy of the President's Counterterrorism Strategy."

[16] Ibid.

Once the target has been approved, a strike is executed only "if we have a high degree of confidence that the individual being targeted is indeed the terrorist we are pursuing," if there is a "near certainty that no civilians will be killed or injured," and if the country where the strike would take place either consents or is unwilling or unable to address the threat.[17] After the strike, forces use "the full range of . . . intelligence capabilities" to determine if the target was killed as well as any collateral damage.[18] If innocent civilians were harmed, Brennan notes that U.S. forces review their actions and strive to make improvements. Moreover, operating forces "regularly" report to Congress and congressional committees on strikes carried out.[19]

LEGAL ISSUES

There is significant controversy in the international community over the legality of the United States' CT campaign. Some of the issues of contention might be alleviated by further clarification of the government's legal position, though one military attorney noted the current lack of consensus around even the basic framework of the law:

> . . . [The] overt disagreement on the answer to the legality question masks that the various participants in the discussion are utilizing wholesale different methodologies and talking past each other in the process. Some speak in terms of how the United Nations Charter governs the overarching question of legality; others claim that the Charter provides only some of the framework; and still others posit that the Charter does not meaningfully apply at all. This divergence leads to correspondingly varied answers as to what extent the law of armed conflict (LOAC) or human rights

---

[17] See, respectively, ibid.; Obama, "Remarks by the President at the National Defense University;" Brennan, "The Ethics and Efficacy of the President's Counterterrorism Strategy"; and Holder, "Attorney General Eric Holder Speaks at Northwestern University School of Law."

[18] Brennan, "The Ethics and Efficacy of the President's Counterterrorism Strategy."

[19] Ibid.

law applies to the use of force through the United States engaging targets in Pakistan. These answers range from the characterization of the conflict in Pakistan as a war and UAS [Unmanned Aerial System] strikes as "just the killing of the enemy, wherever and however found" to the same strike being labeled extrajudicial killings, targeted assassination, and outright murder.[20]

A UN special rapporteur noted that the United States' use of drone strikes outside areas of active hostilities "gives rise to a number of issues on which there is either no clear international consensus, or United States policy appears to challenge established norms."[21] Even former CIA Director Hayden reportedly said that "virtually nobody in the rest of the world agrees with [the] United States targeting policy."[22] Moreover the UN special rapporteur noted that international consensus is lacking on a number of legal issues of importance to drone strike operations, and urged the United States (and other member states) to "further clarify its position."[23]

One of the primary issues revolves around the ambiguity and subjectivity in the definition of an "armed conflict." As mentioned above, some contend that a portion of the areas in which U.S. CT operations, such as drone strikes, take place should be considered below the "threshold" of armed conflict and that IHRL targeting standards should be used.

Another issue of controversy centers on the government's criteria to satisfy a target posing an imminent threat, which have been criticized for being overly broad. Indeed, a DOJ legal memo argues that, due to the nature of terrorism, "Delaying action against individuals continually planning to kill Americans until some theoretical end stage of the planning for

---

[20] Chris Jenks, "Law from Above: Unmanned Aerial Systems, Use of Force, and the Law of Armed Conflict," *North Dakota Law Review* 85 (12 March 2010): 651, http://papers.ssrn.com/sol3/papers.cfm?abstract_id=1569904.

[21] Emmerson, *Promotion and Protection of Human Rights*.

[22] See "Drone Wars," video, 1:29:34, from David M. Kennedy Center for International Studies panel discussion on Simon Schorno, "Drones Wars," *Intercross* (blog), ICRC, 11 March 2013, http://intercrossblog.icrc.org/blog/drones-wars.

[23] Emmerson, *Promotion and Protection of Human Rights*.

a particular plot would create an unacceptably high risk that the action would fail and that American casualties would result."[24] It concludes that, "the condition that an operational leader present [sic] an 'imminent' threat of violent attack against the United States does not require the United States to have clear evidence that a specific attack on U.S. persons and interests will take place in the immediate future."[25] While the DOJ makes a crucial point, this usage is at odds with the definition of the word "imminent" and signals a departure from the historical criteria for a nation to act preemptively in self-defense only when the "necessity of that self-defense is instant, over-whelming, and leaving no choice of means, and no moment of deliberation."[26] Incidentally, targeting under IHRL requires that imminence be established.

Note that this example serves as a caution for the policy options that entail releasing further information about the drone strike targeting processes and the targets themselves. While such releases may be intended to bolster U.S. legitimacy in the eyes of the public and the international community, they could backfire if U.S. practices are confusing, appear questionable, or do not stand up to scrutiny.

These types of ambiguities and controversies seem to have resulted in confusion in the public over what legal framework the government is using, although the government has specified its use of the armed conflict framework. The government's policy of having especially "high and rigorous standards" for targeting beyond what is required in IHL may add to the confusion, for along with the government's emphasis on requiring an "imminent" threat, a typical expectation of no innocent casualties is also reminiscent of the targeting standards from IHRL, but is not a requirement

---

[24] *Lawfulness of a Lethal Operation Directed Against a U.S. Citizen Who Is a Senior Operational Leader of Al-Qa'ida or an Associated Force* (Washington, DC: Department of Justice, 2011), 7, http://www.justice.gov/oip/docs/dept-white-paper.pdf.

[25] Ibid.

[26] Anthony Clark Arend, "International Law and the Preemptive Use of Military Force," *Washington Quarterly* 26, no. 2 (2003): 89–103, http://www.cfr.org/content/publications/attachments/highlight/03spring_arend.pdf.

of IHL.[27] The growing conflation of concepts from IACs and NIACs[28] and the conflation of the legal frameworks governing armed conflicts and terrorism also add to the confusion.[29]

Indeed, acting in a manner consistent with international law while combatting this new type of terrorist threat is not easy. As one report notes:

> The rise of transnational non-state terrorist organizations confounds preexisting legal categories. In a conflict so sporadic and protean, the process of determining where and when the law of armed conflict applies, who should be considered a combatant and what count as "hostilities" is inevitably fraught with difficulty. . . . The legal norms governing armed conflicts and the use of force look clear on paper, but the changing nature of modern conflicts and security threats has rendered them almost incoherent in practice. Basic categories such as "battlefield," "combatant" and "hostilities" no longer have clear or stable meaning.[30]

The way the United States is waging the present "transnational NIAC" has novel elements compared to how NIACs were fought in the past. History has yet to show whether this represents a paradigm shift and adaptation of international law or a deviation from the law.

One example of the novel aspects of the current U.S. framework is the level of participation in hostilities by nonmilitary U.S. personnel, such as OGA personnel and military contractors carrying out missions, such as drone strikes, as well as the reported military support to actions under the

---

[27] The government's policy of "high and rigorous standards" is specified in Brennan, "The Ethics and Efficacy of the President's Counterterrorism Strategy."

[28] See, for example, Michael W. Lewis, "Michael Lewis' Response to Gabor Rona on Targeted Killing," *Opinio Juris* (blog), 1 August 2012, http://opiniojuris.org/2012/08/01/michael-lewis-response-to-gabor-rona-on-targeted-killing/.

[29] *International Humanitarian Law and the Challenges of Contemporary Armed Conflicts.*

[30] Gen John P. Abizaid and Rosa Brooks, *Recommendations and Report of the Task Force on US Drone Policy* (Washington, DC: Stimson Center, 2014), http://www.stimson.org/spotlight/recommendations-and-report-of-the-stimson-task-force-on-us-drone-policy/.

purview of OGAs.[31] This raises questions about whether the protections IHL affords to combatants (combatant privilege and prisoner of war [POW] status if captured) apply to these individuals.[32] In the specific case of drone strike operations, however, the questions may be primarily academic as many U.S. personnel involved in drone strikes operate from within the United States and, in any case, al-Qaeda has been known to have brutally killed U.S. and Coalition forces in Afghanistan without regard for IHL. However, this suggests that the military preference for drone strike operations might be preferable to an absence of such a policy from the standpoint of international law, or the spirit thereof.

On a related note, questions have also been raised about whether military servicemembers involved in covert actions under Title 50 would be entitled to the protections afforded to combatants by IHL.[33] Concerns along these lines would be best addressed by restrictions on DOD conducting covert actions—such as the military preference policy with the additional preference that actions be under Title 10—without stricter limitations on OGAs, if indeed the military carries out covert actions and there are reasons to think that current restrictions and processes are insufficient. Appendix B notes that operating under Title 10 does not restrict the military from carrying out unacknowledged traditional military activities (TMAs), so this option might not sufficiently address this issue.

Exploring further the differences between DOD and OGAs that are relevant to IHL shows that DOD is explicitly obligated to comply with IHL; and its tactics, techniques, and procedures (TTPs) must be in accordance with international law. On the other hand, whether or to what extent OGAs are in practice bound by IHL and other international laws in their

---

[31] For example, see Eric Schmitt and Mark Mazzetti, "Secret Order Lets U.S. Raid Al Qaeda," *New York Times*, 9 November 2008, http://www.nytimes.com/2008/11/10/washington/10military.html?pagewanted=all&_r=2&.

[32] See Joseph B. Berger III, "Covert Action: Title 10, Title 50, and the Chain of Command," *Joint Force Quarterly* 67 (2012).

[33] Marshall Curtis Erwin, *Covert Action: Legislative Background and Possible Policy Questions* (Washington, DC: Congressional Research Service, 2013), http://www.fas.org/sgp/crs/intel/RL33715.pdf.

actions remains unclear.[34] A further difference between DOD and OGAs is the relatively high level of transparency in DOD's chain of command. For these reasons, the military preference has the potential to provide more confidence in the legality of U.S. drone strike practices.

Another notable aspect of the current U.S. framework is the lack of limitation on the geographic scope of the U.S. campaign against al-Qaeda, as reflected in the AUMF. Whether this is appropriate and legal is another point of contention in the international community.[35] What is not under debate is the reality that legitimate al-Qaeda–related threats have operated in multiple areas around the globe. The U.S. government attempts to assuage international concerns with its policy to respect nations' sovereignty and act in a country against a threat only with that country's consent or if it is unable or unwilling to effectively act against the threat.[36]

Another issue to consider is that three of the government's stated standards of targeting—an individuals' membership in al-Qaeda or an associated force, the individuals representing an imminent threat to U.S. persons, and a near certainty of no civilian casualties—appear to be at odds with the widely reported U.S. practice of "signature" drone strikes in which unknown individuals are targeted based on their patterns of behavior. Indeed, this practice is highly controversial, both within and outside of the United States.[37] In combat zones, there is no requirement in IHL to know the identity of targeted individuals, so the concern lies in carrying out signature strikes outside areas of active hostilities, and is one of the controversial consequences of the geographic scope of the U.S. campaign and its

---

[34] Note that this uncertainty is probably by design as some OGA tactics may be more effective if the limits of their actions are not known.

[35] See, for example, Emmerson, *Promotion and Protection of Human Rights*.

[36] See Brennan, "The Ethics and Efficacy of the President's Counterterrorism Strategy"; and Holder, "Attorney General Eric Holder Speaks at Northwestern University School of Law."

[37] See Greenfield, "The Case Against Drone Strikes on People Who Only 'Act' Like Terrorists"; and "Letter from Congressmen John Conyers Jr., Jerrold Nadler, and Robert C. Scott to the Honorable Eric Holder," 21 May 2012, https://www.propublica.org/documents/item/605043-conyers-nadler-scott120521. Also see the minority views in *Intelligence Authorization Act for Fiscal Year 2014*, H. R. 3381, 113th Cong. (2014), https://www.govtrack.us/congress/bills/113/hr3381.

classification as an armed conflict by the United States.[38] Some opponents have called for an end to signature strikes.[39] Short of that, releasing information on the general parameters for these strikes—either by providing more details about the targeting process or about the targeted individuals—together with the legal rationale for these strikes could assuage some of this controversy, although if unconvincing, then such releases would just confirm concerns about signature strikes.

One final issue, noted by a UN special rapporteur, is that in an armed conflict, "in any case in which civilians have been, or appear to have been, killed, the State responsible is under an obligation to conduct a prompt, independent and impartial fact-finding inquiry and to provide a detailed public explanation."[40] This level of transparency to the public—and accountability, depending on DOD and OGA internal practices for holding inquiries—is not currently present in the system. Instituting an Israeli-style court to review instances of civilian deaths could provide a good mechanism for carrying out such investigations. These issues are revisited next, as one component of the larger concept of legitimacy.

LEGITIMACY ISSUES

A number of controversies about the legitimacy—to Western, international, and American audiences—of U.S. drone strike operations have been raised in the public sphere by policy proponents, academicians, defense analysts, lawmakers, and pundits. The issues around legality discussed above are among them, as legality contributes significantly to the legitimacy of these operations. The other issues raised fall into the categories of transparency, accountability, and ethical standing; however, secrecy around U.S. actions also plays into this topic. Select issues are discussed in this section, to-

---

[38] Signature strikes are a complex topic of much importance, however, an analysis of the practice—from a legal perspective or otherwise—is outside the scope of this work.

[39] For example, Micah Zenko, *Reforming U.S. Drone Strike Policies*, Council Special Report No. 65 (New York: Council on Foreign Relations/Center for Preventive Action, 2013), http://www.cfr.org/wars-and-warfare/reforming-us-drone-strike-policies/p29736.

[40] Emmerson, *Promotion and Protection of Human Rights*.

gether with how (or if) they could be addressed by the drone strike policy options.

When it comes to addressing public controversies over drone strikes (or any other practice), note that it is the perception of legitimacy to the public more than legitimacy itself that will be effective. In other words, legitimate practices that lack public visibility will not mitigate public controversies, while effectively hiding illegitimate practices will keep public controversies from worsening. This latter reality is discussed further in the section below on secrecy. At the same time, note that increased transparency (and publicly outlining positive practices when their details cannot be released) can mitigate the former point.

Legality
The previous section outlined a number of issues for drone strikes related to international law. Those issues are summarized below. Concerns and controversies with regard to domestic law exist but are outside the scope of this work.

Controversy Over Specific Aspects of the United State's Legal Rationale
Legal experts have challenged certain specific aspects of the U.S.'s legal rationale. Issues include:

- whether an armed conflict legal framework should be used in all areas where the government is conducting drone strikes and other such operations, as the United States has put forth;

- whether the government's concept of an individual posing an "imminent" threat is too broad; and

- whether the geographic scope of the current CT campaign should be unlimited, as the United States argues.

While significant areas of controversy, these issues are not addressed by any of the policy options considered here.

Questions about IHL Protections to Nonmilitary and Military Personnel
Some military personnel have raised questions about to what extent IHL protections apply to:
- nonmilitary personnel involved in drone strike operations; and[41]
- military personnel carrying out covert actions.[42]

Assuming that IHL protections do not completely carry through in these settings, note that the military preference policy could help address the first issue and the military preference policy with a Title 10 preference could help address the second. This suggests that the military preference policy may be more consistent with the spirit of some aspects of international law than current U.S. practices.

Lack of OGA Transparency
At present, the extent to which OGAs comply with international law remains unclear. Furthermore, there is little transparency in operational OGA chains of command. The military does not have these issues, thus the military preference policy would address these concerns in the context of drone strikes.

The Legality of Signature Strikes and Their Consistency with Stated Targeting Policies
Signature strikes are highly controversial and appear not to necessarily meet certain stated thresholds for targeting: that the target is a member of al-Qaeda or an associated force, that the target represents an imminent threat, or a near certainty of no civilian casualties. Releasing information about targeting processes/standards or about targeted individuals could help to mitigate this controversy.

---

[41] Berger, "Covert Action."

[42] Erwin, *Covert Action*.

The Potential Obligation for Inquiries into and Public Explanation of Civilian Casualties

A UN special rapporteur has asserted that countries have an obligation within armed conflict to run an inquiry and provide detailed public explanation whenever civilians have been killed (or appear to have been killed).[43] An Israeli-style drone court could provide a mechanism for this.

Transparency

Significant Information about Targeting Policy and Practices Are Unknown

The section that introduced the two policy options for releasing more information about targeting processes/standards and targeted individuals highlighted a number of important questions about drone strikes that have thus far gone unanswered by the government. Providing at least some of this information would greatly increase the transparency of drone strike operations.

In addition, the military preference policy would reduce the role of OGAs in drone strike operations. Since so little information is publicly available about OGA drone strike practices, these policies would also increase transparency.

Oversight by Congressional Committees Is Largely Classified

Most oversight (such as hearings) of the House and Senate Intelligence Committees over drone strike operations is classified and, as a result, is not publicly available. In addition, most of the House and Senate Armed Services Committees' oversight related to DOD drone strikes is also classified. While this may be necessary, it would increase transparency if more oversight were conducted at the unclassified level and publicly released.

Certain committees might be able to more effectively declassify and release oversight products; in this case, the military preference policy would have an implication for this issue. If a preference for Title 10 drone

---

[43] Emmerson, *Promotion and Protection of Human Rights*.

strikes were a part of the military preference, fewer covert drone strikes might be carried out and hence there would be less secrecy around drone strikes overall. Strike information and oversight may still be classified, but it would be less sensitive and perhaps more suitable for public release.

An Israeli-style drone court could be used as a tool to publicly release more oversight information. A FISA-like court might have an indirect, though perhaps marginal, effect on this issue, as the court could take some of the pressure for oversight off of Congress. In that case, Congress might regard some of its oversight proceedings as less politically sensitive and therefore would be more inclined to release additional information from its proceedings.

Finally, releasing additional targeting information would be consistent with a push to declassify and release drone strike information, and might allow the committees to release more information about their work.

Supporting and Supported Roles within an Operation Blur Legal Distinctions
As discussed earlier, the DOD may act in support of an operation that an OGA has the "lead" on. In fact, this type of relationship is not unusual within some parts of the government (and within the military itself). For example, a CIA specialist could be temporarily assigned to an FBI team to support certain domestic operations that the CIA would not be authorized to carry out, or military special operations forces could augment a CIA team in a covert operation under CIA command and control and under Title 50 authority.

A more extreme example would be when an entire unit from one organization is placed under tactical control of a separate "lead" agency, and acts with more minimal involvement from the lead agency, although the lead agency bears ultimate responsibility for the operation. This appears to have been the case for the Abbottabad raid that killed Osama bin Laden in 2011. In his interview on CNN following the raid, then-CIA Director Panetta stated that he commanded the operation, but that "the real commander was Admiral McRaven because . . . he was actually in charge of the military operation that went in and got bin Laden."[44]

These types of situations—perhaps unsurprisingly—can lead to public confusion around issues of authority, although authority and chain of command may be clear to the operators carrying out the missions.[45] Public visibility (and perhaps even visibility from within the government) into these types of operations can be lacking, and carrying out this type of support could potentially enable organizations to skirt certain oversight mechanisms.

The military preference policy might make these types of situations rarer by imposing a partiality for the military to be the lead on drone strike operations and therefore not be subordinate to an OGA, although it might also increase the likelihood that OGA forces would operate subordinate to DOD. Personnel currently involved in drone strike operations should be able to analyze the extent to which these types of lead/supporting relationships are present during the operations, and how much the policies would increase or decrease the practice. Releasing additional details about targeting processes could also clarify this practice.

Accountability

Two issues are presented in this section on the topic of accountability, both of which focus on strike casualties.

Information Not Released After Strikes

There are those who argue that the lack of publicly released information on casualties from individual drone strikes—both targets and collateral damage—creates an "accountability vacuum." One report asserts that, "We do not believe it is consistent with American values for the United States to carry on a broad, multi-year program of targeted strikes

---

[44] "CIA Chief Panetta."

[45] It has been contended, however, that Panetta's description of the operation illustrates that "critical confusion exists even among the most senior U.S. leaders about the chain of command and the appropriate classification of such an operation." Berger, "Covert Action."

in which the United States has acknowledged only the deaths of four U.S. citizens, despite clear evidence that several thousand others have also been killed."[46]

Releasing information on collateral damage casualties as well as successfully targeted individuals could bolster U.S. accountability, transparency, and credibility. Obviously the policy option of releasing details about targeted individuals would be a component of this; an Israeli-style drone court could provide a more complete mechanism that covers collateral damage casualties as well.

Inadequate U.S. Government Civilian Casualty Reporting
As discussed above, the U.S. government has no process for publicly reporting civilian casualty estimates resulting from its drone strike operations. More generally, the government has never released any comprehensive data on these events. Independent estimates of civilian casualties are significantly larger than the sporadic U.S. government claims of casualties.[47] Moreover, these U.S. government claims appear to lack credibility, with official statements referring to few or zero casualties having been discredited in some instances.[48] All this detracts significantly from the perceived legitimacy of the U.S. CT campaign. This issue is reflected in the report of the UN special rapporteur on the promotion and protection of human rights and fundamental freedoms while countering terrorism, which specifically called on the United States to release its civilian casualty estimates.[49]

---

[46] Abizaid and Brooks, *Recommendations and Report of the Task Force on US Drone Policy*.

[47] See part 1 of this work, "Drone Strikes in Pakistan: Reasons to Assess Civilian Casualties."

[48] For statements of few casualties, see, for example, Lee Ferran, "Intel Chair: Civilian Drone Casualties in 'Single Digits' Year-to-Year," ABC News, 7 February 2013, http://abcnews.go.com/blogs/headlines/2013/02/intel-chair-civilian-drone-casualties-in-single-digits-year-to-year/; and Scott Shane, "CIA Is Disputed on Civilian Toll in Drone Strikes," *New York Times*, 11 August 2011, http://www.nytimes.com/2011/08/12/world/asia/12drones.html?pagewanted=all.

[49] Emmerson, *Promotion and Protection of Human Rights*.

As with the previous issue, an Israeli-style drone court could provide a mechanism for more consistent public reporting of civilian casualties, depending on how the court was set up.

Ethical Considerations
Civilian Casualties, Apologies, and Redress
Drone strikes cause civilian casualties, both as collateral damage and in the case of civilians being misidentified as al-Qaeda affiliates.[50] This is an ethical issue as well as one of TME. The casualties may not meet the threshold of proportionality; in this instance, due diligence may not have been performed. In either of these cases, ethical questions related to the loss of life remain, as does an onus to minimize civilian harm going forward. The framework described above on the strike process and operational considerations will provide one way to predict the effect of the policy options on civilian casualties.

Aside from the existence of civilian casualties, questions have been raised about whether the U.S. government reacts adequately when civilian casualties occur, both in its acknowledgement of and apologies for them, and also about whether victims and families of victims are able to seek redress. The government established effective practices to apologize for accidental civilian deaths and provide reparations in Afghanistan, although these are not required under IHL.[51] An Israeli-style drone court could provide a streamlined means for doing this in the future.

Stress on Populations in Operating Areas
In addition to the toll taken on victims of drone strikes and their families and friends, drone strikes can be traumatic for local populations. Living underneath armed drone operations, the local public can come to feel a constant fear of attack, augmented by a buzzing sound day and night in places

---

[50] See part 1 of this work, "Drone Strikes in Pakistan."
[51] Ibid.

where the drones are audible.[52] Beyond the ethical issues at play, this stress on the local population risks increasing radicalism and anti-Americanism. Revealing more details about U.S. targeting policy and those suspects who have been targeted (the latter potentially through an Israeli-style drone court) could prevent the local population from seeing the strikes as random. It could also give the population greater control over their own fates by giving them ways to avoid being mistakenly targeted or becoming collateral damage, although obviously terrorists would be privy to this information too. Decreasing the number of drone strikes may offer the biggest mitigation for this problem, however, regardless of whether it results from the military preference policy, a FISA-like drone court, or indirectly (via a chilling effect) from an Israeli-style drone court.

Miscellaneous Issues

Room to Improve: Setting Drone Strike Precedence
One think tank reported that more than 70 countries use drones, though only a small minority operate armed ones.[53] China reportedly considered using a drone strike to kill a drug lord in Burma (but captured and tried him instead, perhaps to avoid some of the controversies discussed herein), and is putting significant investment into drone technologies, including

---

[52] See *The Civilian Impact of Drones*; and Jonathan Cook, "Gaza: Life and Death under Israel's Drones," *Al Jazeera*, 28 November 2013, http://www.aljazeera.com/indepth/features/2013/11/gaza-life-death-under-israel-drones-20131125124214350423.html.

[53] The think tank report is cited in, Peter Bergen, "Drone Wars: The Constitutional and Counterterrorism Implications of Targeted Killing" (written testimony, U.S. Senate Committee on the Judiciary, Subcommittee on the Constitution, Civil Rights and Human Rights, 23 April 2013), http://www.judiciary.senate.gov/imo/media/doc/04-23-13BergenTestimony.pdf. Regarding the operation of armed drones see, for example, Weizmann, "Remotely Piloted Aircraft and International Law"; and "Reaper MQ9A RPAS," Royal Air Force, accessed 27 June 2014, http://www.raf.mod.uk/equipment/reaper.cfm.

its first stealth drone.[54] Moreover, the commercial use of drones sits on the horizon.[55]

The United States is currently setting precedents for drone use, particularly for the purpose of targeted killing. Of the numerous other countries investing in drones, some do not share America's values and interest in complying with international law, and the United States might not be satisfied with those countries carrying out drone strikes with the same level of transparency and so on that it practices.[56] Thus, it is all the more important to develop responsible standards for drone use at this early stage, when U.S. influence is likely maximized. As CIA Director Brennan noted, "If we want other nations to use these technologies responsibly, we must use them responsibly. If we want other nations to adhere to high and rigorous standards for their use, then we must do so as well."[57] If the government firmly established high standards for the use of armed drones then, even if a rogue country did not adhere to the standards, the international security situation would likely be improved as other nations would be more likely to rally against the rogue nation.

---

[54] Jane Perlez, "Chinese Plan to Kill Drug Lord with Drone Highlights Military Advances," *New York Times*, 20 February 2013, http://www.nytimes.com/2013/02/21/world/asia/chinese-plan-to-use-drone-highlights-military-advances.html?_r=0; and Phil Stewart, "Chinese Military Spending Exceeds $145 Billion, Drones Advanced: U.S." Reuters, 6 June 2014, http://www.reuters.com/article/2014/06/06/us-usa-china-military-idUSKBN0EG2XK20140606.

[55] For example, Dara Kerr, "Amazon Delivery Drones Edge Closer to Reality," CNET, 10 April 2014, http://www.cnet.com/news/amazon-delivery-drones-edge-closer-to-reality/; Julianne Pepitone, "Domino's Tests Drone Pizza Delivery," CNN Money, 4 June 2013, http://money.cnn.com/2013/06/04/technology/innovation/dominos-pizza-drone/; "Award-winning Scots Bakery Set to Use Unmanned Drones to Deliver Sweet Treats to Customers," *Daily Record*, 1 April 2014, http://www.dailyrecord.co.uk/news/scottish-news/award-winning-scots-bakery-set-use-3320052; and Ryan W. Neal, "Drunk Drones: Minnesota Brewery Makes World's First Beer Delivery with Unmanned Aircraft," *International Business Times* (video), 17 March 2014, http://www.ibtimes.com/drunk-drones-minnesota-brewery-makes-worlds-first-beer-delivery-unmanned-aircraft-video-1561881.

[56] John P. Abizaid and Rosa Brooks, "U.S. Should Take Lead on Setting Global Norms for Drone Strikes," *Washington Post*, 26 June 2014, http://www.washingtonpost.com/opinions/us-should-take-lead-on-establishing-global-norms-for-drone-strikes/2014/06/25/8183e7ea-fb0b-11e3-b1f4-8e77c632c07b_story.html.

[57] Brennan, "The Ethics and Efficacy of the President's Counterterrorism Strategy."

All of the policy options considered here have the potential to increase the standards for legality, transparency, accountability, and/or ethics in executing drone strikes and therefore to set a more stringent precedent. In particular, an Israeli-style court that evaluates strikes and the grievances of victims and their families, and increasing transparency with targeting policies and past strikes are options that could set especially positive and consequential procedural and legal precedents for drone usage.[58]

Drone Strikes Are Inherently a Military Activity
Implicit in many of the writings about U.S. drone strike operations is that these operations are inherently a military activity, and so it is proper that they be undertaken by the military. Indeed, they are disciplined lethal operations carried out using military weapons within what the George W. Bush administration termed the "Global War on Terrorism." The military preference policy would better align drone strike operations with this perspective.

Secrecy
The extent to which the United States can engage in such operations as drone strikes clandestinely or covertly is to some extent an issue of military effectiveness, and has implications for the perception of U.S. legitimacy. If drone strikes were largely kept a secret from the international public or from those in the country in which the operations take place, this could support the perception that the U.S. government respects state sovereignty, which might increase the perception of U.S. legitimacy. Moreover, these actions may effectively help the United States achieve its CT objectives without a large ground operation, thus avoiding what might be seen as a more illegitimate action than a drone strike operation. Hence, effec-

---

[58] One report recommends going much further in this vein than the listed policy options, urging the United States to "foster the development of appropriate international norms for the use of lethal force outside traditional battlefields." See Abizaid and Brooks, *Recommendations and Report of the Task Force on U.S. Drone Policy*.

tively maintaining the secrecy of its operations can be a tool with which the United States increases or maintains the perception of its legitimacy.

As noted in Appendix A, even if total secrecy is not attained but the strikes are not widely known, the host country may choose not to acknowledge them for diplomatic or practical reasons and, in essence, the "secret is safe" from the perspective of the broader international community.

This practice carries some risk, however, because if the actions done in secret are discovered, the backlash and perception of legitimacy could be more negative than if the operation had been done in the open. Moreover, with respect to drone strike operations specifically, strikes obviously leave evidence that they occurred (see Appendix A for further discussion). Given that very few nations in the world currently operate armed drones, the ability to attain true secrecy may be questionable.

The military preference policy, especially if it includes a preference for Title 10 action, could decrease the flexibility the United States has to carry out secret drone strike operations. This outcome thus could have both potential risks and rewards with regard to the perception of U.S. legitimacy.

SUMMARY OF LEGITIMACY ISSUES

The perception of legitimacy of U.S. drone strike operations is hindered by a number of issues related to legality, transparency, accountability, ethics, and secrecy. Changes in drone strike policy can address these issues to varying degrees.

Of the policy options considered here, an Israeli-style drone court and releasing additional detail about targeted individuals would address the greatest number of issues, while a FISA-like drone court would address the fewest. At the same time, all of the issues listed above could be significantly addressed by at least one of the policy options considered here except for the controversies over the U.S.'s legal rationale, the ethical considerations of civilian casualties, and perhaps the stress drone strike operations cause on local populations.

CHAPTER 11

ANTICIPATING NET EFFECTIVENESS

The Obama administration has noted that the current CT campaign "is an effort to dismantle a specific group of networks that pose a threat to the United States. . . . You cannot eliminate terrorism."[1] This speaks to an end goal of keeping the United States and its citizens largely safe from terrorist attack. This chapter outlines a general way to use the considerations raised on the previous pages to anticipate how changes in drone-strike policy will contribute to or detract from this goal, which we refer to as "net effectiveness."[2]

TME and the perception of legitimacy have direct impacts on net effectiveness.[3] For instance, greater TME might mean that more al-Qaeda leaders are killed and hence those individuals never carry out any planned terrorist attacks; and a greater perception of legitimacy might mean that

---

[1] "Background Briefing."

[2] The notion of "net effectiveness" could be broadened considerably; arguably any U.S. actions abroad have both an immediate goal (such as using CT operations to increase safety from terrorist attacks) as well as goals of maintaining or furthering Western and American values and interests. A more extensive analysis along the lines of what is presented here could incorporate these much broader goals.

[3] A similar point to this—and the further discussion in the paragraphs below—could be made for OME and SME just as well as for TME. Here as in the rest of this work, however, the focus is restricted primarily to TME.

anti-American sentiment would decline and fewer people would be motivated to perpetrate terrorist attacks against the nation and its interests.

TME and legitimacy are also interrelated. They can bolster one another; for example, a greater perception of legitimacy can result in increased operational support from allies, such as increased intelligence sharing and allowing the use of their airspace, which could lead to more accurate and timely targeting, among other things. At the same time, improvements to targeting and operating procedures could mean that fewer civilian casualties accrue, which bolsters U.S. legitimacy.

On the other hand, TME and legitimacy can also be at odds with one another. For instance, releasing certain specific details about U.S. standards for targeting may add greatly to the transparency of the process but allow terrorists to avoid drone strikes. Alternatively, the U.S. government might achieve greater military success by not taking steps to avoid civilian casualties and other collateral damage, but it would come at the cost of the nation's ethical standing and adherence to international law.

Consider that a successful foreign-based terrorist attack entails:[4]

- the individual or group of terrorists being personally motivated to carry out the attack;

- a failure to stop the attack by the United States and the international community; and

- a failure to stop the attack by the nation from which it was based.

The discussion that follows explores each of these three aspects, and considers how increased perceived legitimacy and TME of U.S. drone strike operations could affect each. While numerous secondary effects could stem from such increases, the discussion below attempts to focus on those that are relatively immediate. Policymakers and other interested parties are en-

---

[4] Due to our focus on U.S. drone strikes abroad, the scope of this discussion is limited to terrorists who plan or launch attacks from abroad vice "homegrown" terror. More specifically, in accordance with stated U.S. policy discussed above, the focus is on only foreign-based attacks from nations that consent to U.S. operations or are unable or unwilling to act against the given threat to the United States.

couraged to consider deeper analysis along these lines in the context of the types of specific scenarios most relevant to their situations.

The individual or group of terrorists being personally motivated to carry out the attack. Disenfranchisement and other factors are linked to radicalization, in general, while negative perceptions of the United States and the West lead to anti-Americanism specifically.[5] Negative perceptions of the United States and the West can be increased by the perception that the U.S. CT campaign is illegitimate; alternatively, they can be decreased by the perceived legitimacy of U.S. actions. Anti-American sentiments also can result from poor military and security effects, such as civilian casualties, or terrorist leaders recruiting and encouraging terrorist action without disruption by military forces. Thus, the perceived legitimacy and TME of U.S. drone strikes (or the lack thereof) could affect the personal motivations of would-be terrorists—both in the local area that strikes take place and more widely around the world.

Failure to stop the attack by the United States and the international community. Increased TME and/or sufficient intelligence might contribute to the United States and the international community being better able to stop a terrorist attack. Obviously, the greater the TME, the more likely that drone strikes would successfully prosecute their targets and thus disrupt terror plots, and the less likely that individuals who further U.S. security (e.g., local leaders) would be mistakenly killed. Increased TME in areas such as these and perceived legitimacy could garner greater support from the local population, making it more inclined to support intelligence collection and actions by U.S. forces (e.g., by allowing greater freedom of action for U.S. forces). Conversely, TME has the potential drawback that successful targeted killings result in a dead end with respect to intelligence collection, since suspects are killed rather than captured and therefore cannot be questioned. Other potential sources of intelligence, such as computers, may also be destroyed by strikes.

---

[5] The link between disenfranchisement and radicalization is discussed in Tori DeAngelis, "Understanding Terrorism," *APA Monitor* 40, no. 10 (November 2009), http://www.apa.org/monitor/2009/11/terrorism.aspx.

The level of support the United States receives from allied and (a priori) neutral nations—which may include the nation where strikes will take place in or third-party nations—would be bolstered by the perceived legitimacy of U.S. operations (and vice versa). Note that such support could include things like sharing intelligence, providing Coalition forces for operations, or allowing U.S. forces to base on their land or transit via their roads, airspace, or territorial waters, all of which could increase military and net effectiveness.

Failure to stop the attack by the nation from which it was based. Contributing factors might include such things as insufficient intelligence, ineffective internal security forces, and/or a lack of political will. The extent to which U.S. actions, and drone strike operations in particular, affect these points would depend on the relationship the country in question—call it the nation containing the target (NCT)—has with the United States. For our purposes, this relationship can be thought of as falling along a continuum that goes from working fully with the United States on CT operations to not cooperating with or even actively working against U.S. interests.

If the NCT cooperates with the United States, then greater legitimacy of U.S. operations could make the local NCT population more sympathetic to the NCT and U.S. campaign, and therefore more inclined to provide accurate intelligence, to support broader freedom of action by NCT security forces, and to give more political support to the NCT for these actions. Political support could further encourage the NCT government to carry out operations against the terrorists in question. In addition, U.S. TME would bolster the ally NCT's TME, if the United States is providing military support to NCT operations. However, if the NCT government is unpopular with the local population, these gains would be diminished and might even be maximized by not publicizing U.S. cooperation with the NCT government.

If the NCT does not cooperate with the United States, then the TME and legitimacy of U.S. drone strike operations would likely have a more minimal effect on the NCT's ability to stop the attack, although the NCT might have more political will to attempt to disrupt the terrorist plot if

U.S. actions abroad are seen as legitimate. If the United States was using drone strike operations against the threat, however, and that practice was viewed as legitimate and resulted in minimal collateral damage, then tensions would not rise as much between the United States and the NCT than would be the case if U.S. strikes were viewed as less legitimate or accrued more collateral damage. Tensions with the NCT could also decrease if U.S. operations were effective and did kill local al-Qaeda leaders and their local political or military constituencies disbanded. There would also be the possibility, however, that the NCT would actively counter U.S. operations. In this case, increased TME could help the United States to defeat these actions. These considerations are summarized in table 2.

Instead of comparing outcomes from U.S. drone strike operations in the NCT to those of U.S. strikes that have a greater perception of legitimacy and are more militarily effective, one can compare these scenarios with other alternatives, such as the United States choosing not to act in

Table 2. Potential advantages of increased TME and perceived legitimacy

Effect of: / Effect on:	Increased U.S. TME	Increased perceived U.S. legitimacy
Individual terrorist	• Anti-American sentiment reduced through fewer civilian casualties and other collateral damage • Radicalization or plot not fully developed due to leader being effectively targeted	• Anti-American sentiment reduced
U.S. and international community's actions against the threat	• More plots disrupted • People who further U.S. security are less likely to be killed • More popular support leading to increased local intelligence and freedom of action	• Increased support from allies • More popular support leading to increased local intelligence and freedom of action
NCT's actions against the threat—ally case	• More popular support leading to increased local intelligence and freedom of action for NCT forces • More political will for NCT to act • Better enable NCT forces	• More popular support leading to increased local intelligence and freedom of action for NCT forces • More political will for NCT to act
NCT's actions against the threat—nonally case	• Reduced escalation between United States and NCT or increased ability to defeat active countering of drones by NCT • More political will for NCT to act	• Reduced escalation between United States and NCT • More political will for NCT to act

the NCT or, at the other extreme, pursuing the threat with ground troops. This would generate a separate calculus with different implications. For example, the use of drone strikes in an NCT that is not an ally would be more escalatory and of greater risk to U.S. security than no action in the NCT, but would likely be less escalatory than sending in ground troops.

In light of this discussion, the following steps are recommended to evaluate the potential net effect of a drone strike policy:

- Using the considerations in the previous sections, determine the extent to which the policy furthers the TME and the perceived legitimacy of drone strike operations;

    - Apply to table 2 (or a more detailed product that focuses on a particular scenario of interest) to yield the extent to which the items in the cells of the table could be expected.

- Use the considerations in the previous sections to determine any risks to TME and perceived legitimacy the policy would entail;

- In a similar fashion, determine OME and SME implications as desired;

- Consider effects of changes in TME and perceived legitimacy, as well as any determined effects on OME and SME, on either bolstering or hindering one another, as discussed earlier in this chapter; and

- Analyze all of these factors in concert to achieve a prediction of the net effect.

Note that carrying out such an analysis of the five policy options considered here will require the details of the intended implementation of the options, as well as input from operators and others familiar with current drone strike practices.

CHAPTER 12

CONCLUSIONS AND RECOMMENDATIONS

The previous chapters presented a framework with which to systematically consider the effects of drone strike policy changes, with a focus on TME, legitimacy, and a methodology for anticipating net effectiveness. This chapter summarizes the findings for each of the policy options considered and provides some concluding remarks for the discussion from Part II.

THE MILITARY PREFERENCE

Instituting a preference that the military perform drone strikes—guidance issued by President Obama in May 2013, but with implementation interrupted as a result of congressional actions—would add transparency and a stronger expectation of legality to drone strike practices since the military's doctrine, operating procedures, and chain of command structure are relatively well understood and are known to be aligned with international law. IHL protections for military personnel also indicate that this option would be better aligned with the spirit of the law.

Ironically, the lack of transparency of the content of the presidential guidance is currently an impediment to analyzing its implications, with the guidance itself classified, as well as the legislation that Congress passed that reportedly limits its implementation. Releasing the content of President Obama's guidance would seem to be in the spirit of the guidance itself, although as with all the other options discussed here, a cost/benefit analysis (using the framework presented above) would be in order.

The military preference policy would have implications on oversight and accountability of drone strikes, as DOD falls under different oversight mechanisms than OGAs. The effects of these implications are not immediately clear, but could be analyzed by those directly involved in drone strike processes. In particular, a comparison of the oversight by the House and Senate Armed Services Committees with that of the House and Senate Intelligence Committees could provide highly relevant insight. The military preference would also likely impact TME—perhaps in both positive and negative ways—and operators and other involved personnel would be in the best position to assess these risks and potential rewards as well.

An evaluation of the military preference policy should focus primarily on determining its anticipated effects on TME, oversight, and accountability. Those effects should be evaluated against the additional transparency and alignment with international law the policy could afford.

A specific preference for drone strikes to be employed under Title 10 could be included as a part of this option.[1] If, in practice, the restrictions this entails effectively lessen the secrecy around drone strike operations (see Appendices A and B), then this option would provide for additional transparency at the cost of more limited options for drone strike operations. Such limitations could increase diplomatic and operational risks.

A FISA-LIKE DRONE COURT

The proponents of a FISA-like court, which would authorize drone strikes, note that from a legal perspective, this practice might be comparable to a court approving a warrant based on probable cause, as is the case for the FISA court.[2] Nonetheless, this option appears to raise the most legal questions (with respect to U.S. law). One report puts forth that "Such a court would likely be unconstitutional because it would violate the separation of powers and would be asked to render advisory opinions rather than

---

[1] Recall that a strike being carried out under Title 10 would mean it was necessarily employed by the military, as a TMA, or otherwise as an action that does not meet the definition of covert given in Title 50.

[2] *Joint Special Operations: Task Force Operations.*

rule on actual cases and controversies. The result would be to give a patina of legitimacy to a ruling for summary execution following a one-sided argument."[3] To those who wonder why a FISA-like drone court would be problematic when a warrant-issuing FISA court is legal, a retired federal judge noted that "the answer is simple: a search warrant is not a death warrant."[4] Moreover, in an armed conflict, a FISA-like drone court would be an intrusion into battlefield decision making and contrary to the chain of command and, outside of an armed conflict, it could be at odds with such requirements as a target being an imminent threat.[5] This again raises the issue of the United States' relatively expansive meaning for "imminent threat."

Legal questions aside, a FISA-like court could bolster the legitimacy of drone strike practices by adding additional oversight to the process, thereby arguably setting a better precedent for the military use of drones. This additional oversight could result in reduced collateral damage and other unintended consequences, and could encourage the government to release some information about individual drone targets after strikes are completed. Those who currently execute drone strikes should use the framework given earlier to determine further tactical and operational effects this option might have, such as whether it would slow down the targeting process and, if so, whether that extra time, together with the additional oversight, would lead to better or worse TME overall.

It is worth noting that the FISA court this policy is modeled after confers only limited legitimacy to the U.S. government's collection of electronic communications. The court is controversial—it reportedly approved all but one of more than 8,000 requests it received from 2009 to 2013, and

---

[3] *How to Ensure that the U.S. Drone Program Does Not Undermine Human Rights: Blueprint for the Next Administration* (New York: Human Rights First, 2012, 2013 rev.), http://www.humanrightsfirst.org/wp-content/uploads/pdf/blueprints2012/HRF_Targeted_Killing_blueprint.pdf.

[4] James Robertson, "Judges Shouldn't Decide about Drone Strikes," *Washington Post*, 15 February 2013, https://www.washingtonpost.com/opinions/judges-shouldnt-decide-about-drone-strikes/2013/02/15/8dcd1c46-778c-11e2-aa12-e6cf1d31106b_story.html.

[5] *How to Ensure that the U.S. Drone Program Does Not Undermine Human Rights*.

is viewed as a rubber stamp by many.[6] If a FISA-like drone court would be expected to behave similarly, these types of controversies might outweigh any increased legitimacy the process would confer.

Overall, however, the primary considerations for a FISA-like drone court appear to be its legality with respect to U.S. law and the overall effect it would have on TME.

AN ISRAELI-STYLE DRONE COURT

An Israeli-style court would review drone strikes after they occur. It could be implemented in a number of different ways, with various options for who oversees the court, to what extent (if any) its findings would be released to the public, and the criteria to determine what strikes it investigates. The court could potentially increase the accountability and transparency of drone strike operations, and could provide a means to communicate targets' terrorist activities to the public or to provide redress to innocent drone strike victims and their families. As discussed above, the UN special rapporteur sees these latter activities as legal requirements.[7]

This option could be used as a mechanism for public reporting of civilian casualty numbers. Risks include the potential for a chilling effect on strike operations in anticipation of this layer of oversight, and the potential public backlash if strikes were found to be done improperly. As with the FISA-like court, it would be worth evaluating the level of legitimacy the Israeli courts confer on drone strikes in Israel to get a sense for the potential pitfalls of this option.

Thus, depending on how Israeli-style court procedures would be implemented, it could play a variety of different roles and bolster U.S. legitimacy to various degrees depending on these roles. The main issues here

---

[6] See, for example, Bryan Denson, "FISA Court, which Approves FBI, NSA Surveillance, Faces Reform Challenge from Oregon Senators," *Oregonian*, 26 November 2013, http://www.oregonlive.com/news/index.ssf/2013/11/fisa-court-oregon-nsa-surveillance.html. The FISA court requests were reported in Colin Schultz, "The FISA Court Has Only Denied an NSA Request Once in the Past 5 Years," *SmartNews*, Smithsonian, 1 May 2014, http://www.smithsonianmag.com/smart-news/fisa-court-has-only-denied-nsa-request-once-past-5-years-180951313/?no-ist.

[7] Emmerson, *Promotion and Protection of Human Rights*.

would be to decide on the optimal implementation of the court based on the most important areas of legitimacy to be addressed and on how well drone strike practices would stand up to various levels of scrutiny, as well as any indirect operational impacts.

RELEASING DETAILS ABOUT TARGETING

The two independent policy options in this category are (1) releasing more information about the United States' targeting processes and standards, and (2) releasing details about targeted individuals after a strike has occurred, as could be done through an Israeli-style drone court or otherwise. Either of these options could significantly increase transparency for drone strike operations, thereby bolstering U.S. credibility, assuming processes are sound and strikes are not found to have been done improperly.

All the policy options considered here have the potential to hinder U.S. military effectiveness through increased transparency, giving adversaries more information about operations that could be used to avoid being targeted. However, for these two policy options, this risk is especially high.

Providing details about targeted individuals would be an unusual measure to take during an armed conflict. Releasing information that harkens to "evidence" might give the impression that drone operations have legal obligations like those of law enforcement, when the U.S. government's position in the current context is that they do not. This situation could raise additional controversy and further confuse perceptions of the legal framework the U.S. government is using.[8]

Some of the current issues surrounding accountability could be addressed by releasing details about targeted individuals. As a caution, however, note that releasing partial information and/or releasing information related only to certain strikes might backfire by drawing more attention to the information that is not released, thereby making the public

---

[8] Perceptions of the United States' legal framework may be confused by the fact that some of the current policies and doctrinal standards surrounding drone strikes already harken to the notion of "evidence," such as President Obama's stated requirement for "near certainty" that civilian casualties be avoided. Obama, "Remarks by the President at the National Defense University."

and the international community even more suspicious of U.S. drone strike secrecy and practices.

In short, these two policy options have the potential to greatly increase the transparency of drone strike operations, but could also potentially entail significant risk to U.S. operations depending on the type of information released. Thus, the specifics of any proposal along the lines of these options should be carefully evaluated and the corresponding trade-offs considered.

OVERVIEW OF INITIAL FINDINGS FOR ALL POLICY OPTIONS

An overview of these findings on the policy options is presented in table 3.

Table 3 shows that the policy options considered have the potential to significantly improve the perception of the legitimacy of drone strike operations, particularly an Israeli-style drone court and releasing information about targeted individuals, but also all potentially pose risks to military effectiveness. However, this is not an "apples to apples" comparison: two cells with the same label do not necessarily indicate identical—or even comparable—effects. Rather, the specific considerations within the body of this work should be taken into account. Furthermore, any effects would also depend on the specifics of the implementation of each policy option. Moreover, the policy options are not mutually exclusive—there are no restrictions against implementing more than one option and, in fact, this may be desirable.

FINAL THOUGHTS

The policy options presented in Part II indicate that there are likely no silver bullets that would bolster (or not hinder) the military effectiveness of drone strikes while staying within the confines of international and domestic law, increasing transparency, and so on. Each of these options addresses some of the current issues and controversies with drone strikes, and yet there are issues that are not addressed by any of them—including unresolved issues related to international law—or are only minimally ad-

Table 3. Potential effects of policy options on military effectiveness and identified legitimacy issues

		Military preference	FISA-like court	Israeli-style court	Release information about targeting process/ standards	Release information about targeted individuals
Impacts on military effectiveness	Tactical strike steps	Effect likely; best assessed by relevant personnel	Effect likely; best assessed by relevant personnel	Effect likely; best assessed by relevant personnel		
	Adversaries' actions	Potential of worsening conditions	Potential of worsening conditions	Potential of worsening conditions	Potential of worsening conditions	Potential of worsening conditions
Addressing legitimacy issues	Legality	Potential improvements		Potential improvements	Potential improvements	Potential improvements
	Transparency	Potential improvements		Potential improvements	Potential improvements	Potential improvements
	Accountability			Potential improvements		Potential improvements
	Ethical considerations			Potential improvements	Limited potential improvements	Limited potential improvements
	Miscellaneous	Potential improvements	Potential improvements	Potential improvements	Potential improvements	Potential improvements
	Issues around secrecy	Effect likely; best assessed by relevant personnel				

Note: potential improvement and potential worsening of conditions are noted, along with areas where an effect is likely but best assessed by DOD, OGA, and other relevant personnel. Blank cells indicate that no effects are noted.

dressed, most notably the ethical considerations of civilian casualties and the effects of drone campaigns on local populations.

In the context of this unfortunate lack of simple solutions, the framework and discussion presented here offer a means with which to consider at least some of the consequential issues related to drone strikes, and provide a tool for determining the implications of proposed changes to drone strike operations.

PART III

IMPROVING LETHAL ACTION

LEARNING AND ADAPTING IN U.S. COUNTERTERRORISM OPERATIONS
BY LARRY LEWIS

CHAPTER 13

INTRODUCTION

The United States uses lethal force to eliminate individuals it believes pose an imminent terrorist threat to its citizens and interests, as well as those of its allies and partners. Such force includes actions taken on the battlefield for declared theaters of conflict, such as Afghanistan. But lethal force has also been used outside areas of active hostilities, such as Yemen, Pakistan, and Somalia. Lethal actions in these particular countries have become much more frequent since 2008, and they are often—though not always—successful in killing the targeted individuals. The use of force, however, can also unintentionally kill civilians. These civilian casualties are tragic and also result in other negative consequences on the ground, such as loss of income for households and stigmatization of civilians mistakenly targeted.[1]

Civilian casualties also reduce the overall effectiveness of the U.S. counterterrorism (CT) effort.[2] This negative effect is the result of multiple factors, including alienating local populations, reducing their willingness to provide intelligence, and creating grievances that can lead to the creation of more terrorists; failing to disrupt the threat if the action did not kill the intended individuals; delegitimizing America's CT efforts in the eyes

---

[1] See *Civilian Impact of Drones*.

[2] For more discussion of this point, see Lewis, *Drone Strikes in Pakistan*.

of directly or indirectly affected foreign populations; and creating political difficulties with our allies and partners.[3]

Given the real potential of negative outcomes from the use of lethal force to undermine U.S. CT efforts, including future uses of lethal force, it would seem prudent for the American government to put in place an effective operations analysis framework and lessons-learned process to ensure that it adapts its CT operations for maximum success. Yet, at least publically, this appears to not be the case. As such, Part III will seek to address this deficiency by presenting an analytic framework and lessons-learned process that the U.S. government should use to continually improve the effectiveness of its lethal force operations and reduce the likelihood of civilian casualties in the future.

BACKGROUND

The Need for Lethal Action

On 11 September 2001, al-Qaeda struck down thousands of innocent civilians on American soil. The United States still faces an ongoing security threat from al-Qaeda and its associated forces, including attacks on U.S. forces and interests on foreign soil and continuing operational planning for further attacks. The government endeavors to address this threat in a number of ways, including working with partner nations to improve their security capacity for addressing al-Qaeda threats within their borders and addressing underlying governance and economic issues that can lead to radicalization and support to terrorist groups. However, sometimes other

---

[3] Senator Elizabeth A. Warren summarized the strategic risks thus: "Do we talk seriously about the price our great nation, built on the foundation of life, liberty, and the pursuit of happiness, may pay if others come to believe that we are indifferent to the deaths of civilians? Do we fully take into account the effect on our interests if people around the world are inflamed by such casualties, or if they do not believe that our actions align with our values?" Senator Elizabeth Warren, "Collateral Damage, National Interests, and the Lessons of a Decade of Conflict" (lecture, Annual Whittington Lecture, Georgetown University, 26 February 2014).

nations are unwilling or unable to address these threats directly. In these cases, the United States can resort to direct action against terror threats.[4]

Direct action includes both capture operations and the use of lethal force (e.g., airstrikes, raids). The former is preferable when feasible; pursuing this option is consistent with both U.S. policy and international law, and is operationally advantageous because of potential intelligence gains.[5] While there are examples of successful U.S. capture operations outside of declared theaters of conflict, several factors impact the feasibility of this option in some situations.[6] As a result, U.S. direct action involving lethal force (subsequently referred to as lethal action) is an important component of the CT campaign outside areas of declared hostilities. For example, the government uses lethal force in CT operations in Yemen and Pakistan, as well as in Iraq against the Islamic State of Iraq and the Levant (ISIL).

Lethal Force Operations: Balancing Objectives
The use of lethal force can have both positive and negative impacts on U.S. objectives in the context of counterterrorism operations—positive impacts from successful missions that kill the intended target, and negative impacts

---

[4] Direct action is defined as "short-duration strikes and other small-scale offensive actions conducted as a special operation in hostile, denied, or diplomatically sensitive environments and which employ specialized military capabilities to seize, destroy, capture, exploit, recover, or damage designated targets." See *Department of Defense Dictionary of Military and Associated Terms*.

[5] Capture also sends a strategic message that terrorists are criminals operating outside of acceptable international norms compared to soldiers in a legitimate conflict. This distinction is consistent with national policy: "We must use the full influence of the United States to delegitimize terrorism and make clear that all acts of terrorism will be viewed in the same light as slavery, piracy, or genocide: behavior that no respectable government can condone or support and all must oppose." *National Strategy for Combatting Terrorism* (Washington, DC: CIA, 2003), 23–24, https://www.cia.gov/news-information/cia-the-war-on-terrorism/Counter_Terrorism_Strategy.pdf.

[6] For example, the 2013 capture of Abu Anas al-Libi in Libya for his role in the 1998 bombings of U.S. embassies in Kenya and Tanzania. The issues of legal custody, sovereignty, and force protection can all complicate or preclude capture operations.

from civilian casualties during these operations.[7] Successful missions and civilian casualties have short- and long-term impacts on the overall CT campaign (table 4).

Table 4. Short- and long-term effects of mission success and CIVCAS

	Mission success	CIVCAS
Short-term effects	Neutralize specific threat Disrupt terror network Buy time for other approaches	Increase popular support to terror networks (insurgent math) Decrease host-nation government support
Long-term effects	Degrade operational capability of terror networks Deny sanctuary Discourage recruiting efforts	Decreased perceived legitimacy and freedom of action Feed long-standing grievances that fuel instability and terror networks

Successful Missions: Impacts

In the near term, effectively dealing with a threat through lethal action can neutralize or delay a specific threat to the United States.[8] For example, in July and August 2013, a series of drone strikes sought to disrupt a terror plot in Yemen targeting Western assets.[9] Besides disrupting specific threats, the strikes also affected the terror network. The impact of this action can vary; in some cases, resulting gaps in terror networks are filled quickly from below, while in other cases the proficiency and/or experience of the individual are not easily filled, resulting in a more significant near-term

---

[7] Other outcomes also can negatively impact lethal action operations, including friendly fire, targeting of host-nation forces or assets, and harming hostages in rescue operations. However, as the most widely discussed negative outcome in national CT guidance is civilian casualties, the treatment of negative impacts here will focus on this issue.

[8] Lethal operations can also have a deterrent effect on host-nation domestic terrorist activity. See, for example, Patrick B. Johnston and Anoop K. Sarbahi, "The Impact of U.S. Drone Strikes on Terrorism in Pakistan and Afghanistan" (unpublished paper, Annual Meetings of the American Political Science Association, the Belfer Center for Science and International Affairs at Harvard University's Kennedy School of Government, and the New America Foundation, 2011), http://patrickjohnston.info/materials/drones.pdf.

[9] Greg Miller, Anne Gearan, and Sudarsan Raghavan, "Obama Administration Authorized Recent Drone Strikes in Yemen," *Washington Post*, 7 August 2013, https://www.washingtonpost.com/world/national-security/us-personnel-evacuated-from-yemen-after-al-qaeda-order-to-attack-is-revealed/2013/08/06/2c984300-fe88-11e2-96a8-d3b921c0924a_story.html.

impact. Conversely, ineffective strikes can complicate operations in the short term by alerting the individual or group being targeted.

In the long term, effective operations can damage the operational capabilities of terrorist networks by removing key individuals with highly valued skills that cannot easily be replaced. They can also deter terrorist activities in targeted areas, complicating recruitment and denying sanctuaries in areas where host nations could or would not act. Ineffective operations can, in the long term, betray intelligence techniques or sources and make follow-on operations more challenging. Also, in some scenarios, the number of opportunities is limited (and becoming even more so over time), so ineffective operations represent significant setbacks. Ineffective strikes also can include cases where the wrong person is inadvertently targeted, which can cause outrage and alienation within the local population.

Civilian Casualties: Impacts
At the same time, negative second-order effects from strikes, such as civilian casualties, can temper or completely blunt the benefits of CT efforts. In the short term, even an effective strike can galvanize the local community to support terrorist networks if the strike results in civilian casualties. The effect, called "insurgent math" by General Stanley McChrystal in Afghanistan, can result in net growth of the terrorist network rather than net attrition.[10] The effect of insurgent math can be magnified in the case of civilian casualties.[11] Overall, civilian casualties from these operations may create grievances that radicalize populations, increase support for terrorist elements, and degrade the political will of the United States and partner

---

[10] The concept of insurgent math is based on the idea that for every innocent person killed, 10 new enemies are created. See Sean D. Naylor, "McChrystal: Civilian Deaths Endanger Mission," *Army Times*, 30 May 2010.

[11] This point was exemplified by Gen McChrystal: "When we fight, if we become focused on destroying the enemy but end up killing Afghan citizens, destroying Afghan property or acting in a way that is perceived as arrogant, we convince the Afghan people that we do not care about them. If we say, 'We are here for you—we respect and want to protect you' while destroying their home, killing their relatives or destroying their crops, it is difficult for them to connect those two concepts. It would be difficult for us to do the same." Gen Stanley McChrystal, "Question and Answer" (remarks, International Institute for Strategic Studies, London, 1 October 2009).

nations for action. In addition, negative second-order effects can reduce freedom of action by decreasing host-nation support, as well as decreasing perceived legitimacy of the United States in the eyes of domestic and international audiences.

In addition to the negative impacts that stem from civilian casualties, negative impacts can also come from false allegations concerning civilian casualties. The allegations may arise from misunderstandings, particularly due to a lack of recognition among the local populace that the targeted individual(s) were in fact terrorists and combatants. However, they can also be deliberate efforts by terrorist groups or others to create a negative perception of the United States and sway certain audiences. Typically, allegations can provide the first impression of the operation, and this incorrect impression can be a lasting one. For example:

- In Afghanistan, after a Taliban improvised explosive device (IED) prematurely detonated in a city center in southern Afghanistan, Taliban elements blamed the explosion on an International Security Assistance Force (ISAF) airstrike. ISAF did not respond effectively to the allegation; as a result, even a year later, locals still believed the story to be true, creating resentment and lack of cooperation with international forces.[12]

- In Iraq, when a U.S. operation against foreign fighters in a safe house was falsely portrayed by terrorists as killing women and children in a wedding party, U.S. forces were slow to respond. To this day, the true nature of the target is widely misunderstood.[13]

Contesting allegations is complicated in covert and clandestine operations, but involvement of the host nation in consequence management—as was done in Afghanistan in later years—is one way to deal with this challenge. The ability to contest such allegations is one reason why it is important

---

[12] Sewall and Lewis, *Joint Civilian Casualty Study*.

[13] *Transition to Sovereignty: Joint Lessons Learned for Operation Iraqi Freedom* (Suffolk, VA: JCOA, 2007).

to have a process for evaluating the presence of civilian casualties during operations.

MANAGING POSITIVE AND NEGATIVE IMPACTS: THE ROLE OF GUIDANCE

The authority to use force carries many responsibilities, such as the creation of guidance and policy to define and govern when force may be used, processes to gain approval for operations, and overall intent for the use of force. The guidance informs the operational approach, including the targeting process, intelligence allocation, requirements for pattern of life determinations, tactics, and training in support of operations.[14]

The United States has refined its guidance regarding lethal force in CT operations over time. The most recent guidance comes from the May 2013 Presidential Policy Guidance (PPG), which focuses on the effective use of lethal force against the most significant security threats to the nation.[15] The PPG aims to ensure that lethal force conforms to U.S. laws, is only used when no other option exists, and meets two primary criteria:

- Operations are effective against the threat; and

- Operations avoid negative second-order effects that can both limit freedom of action and undercut progress against terrorist organizations (e.g., civilian casualties).[16]

As a result of conflicting impacts from effective missions and civilian casualties discussed above, lethal force operations attempt to meet both of these objectives when possible. This is not just true for U.S. CT operations in Pakistan and Yemen. For example, airstrikes against ISIL have focused

---

[14] Pattern of life determinations are the characterization of normal trends of activity, including both suspected threats and civilians.

[15] The PPG is summarized in the unclassified announcement by the White House, Office of the Press Secretary, "Fact Sheet."

[16] Ibid. Other stated criteria in the PPG collectively amount to a requirement that no feasible alternatives exist to effectively address the imminent threat.

on targets away from heavily populated civilian areas to minimize the risk of civilian casualties, and counterinsurgency and counterterrorism operations in Afghanistan also sought to meet these objectives.

Yet guidance for the use of force—no matter how well intentioned—does not always have its desired effect. Afghanistan is an instructive example. When combat operations caused a significant number of civilian casualties in 2007—harming progress in the campaign as well as the relationship with the host nation government—then-Commander, International Security Assistance Force (COMISAF) General Dan K. McNeill issued a tactical directive in June 2007 to reduce civilian casualties in ISAF operations, with goals and guidance similar to what would later be seen in the 2013 PPG.[17] Despite the guidance, however, the intent was not met, with significant civilian casualty incidents occurring throughout 2007 and 2008. Several subsequent tactical directives were released by General David D. McKiernan, but they had a similar—that is, negligible—effect on reducing the level of civilian casualties.[18]

It was not until a revised directive was issued in 2009 that some improvement could be seen in ISAF forces' efforts to reduce civilian casualties. A further revision, issued in 2010, was even more effective. Similarly, Special Operations Forces (SOF) made changes to their operations in 2010–12 that both reduced civilian casualties and improved mission success. What made the difference? Guidance prior to 2009 provided intent but was uninformed with respect to the causal factors that contributed to civilian casualties, which limited the utility of the guid-

---

[17] For more information, see Jennifer Keene, *Civilian Harm Tracking: Analysis of ISAF Efforts in Afghanistan* (Washington, DC: Center for Civilians in Conflict, 2014), 3, http://civiliansinconflict.org/uploads/files/publications/ISAF_Civilian_Harm_Tracking.pdf.

[18] Lewis, *Reducing and Mitigating Civilian Casualties*.

ance.[19] In contrast, two elements made the difference in the improvements for 2009 and 2010. In 2009, leadership emphasized existing guidance. The 2009 tactical directive was not substantively different than previous guidance, but the commander and subordinate echelons heavily emphasized the content, improving consistent implementation and promoting creative problem solving at the tactical level.[20] The biggest improvement occurred in 2010, when General Petraeus implemented a revised tactical directive, which included additional considerations that reflected lessons from early incidents, allowing forces to learn from past experiences rather than repeat mistakes. These results show the benefit of evaluating performance, understanding key drivers of success and failure, and then revising guidance to reflect an improved understanding.

---

[19] In a similar example, U.S. forces struggled with limiting civilian casualties during escalation-of-force incidents. Analysis showed that existing doctrine, and much equipment, was optimized for one root cause of civilian casualties based on experiences in Bosnia, however, the majority of incidents in Afghanistan did not share this root cause. Later guidance reflected this disconnect between guidance and operational reality, and civilian casualties decreased. See Lewis, *Reducing and Mitigating Civilian Casualties*.

[20] Ibid.

CHAPTER 14

AN ANALYTICAL APPROACH TO IMPROVE LETHAL ACTION OPERATIONS

Given the historic limitations of guidance that achieves its stated intent, it is worthwhile to consider how U.S. counterterrorism operations have performed with respect to the goals of the PPG issued in 2013. This chapter presents an analytical approach (figure 5) that answers this question but also identifies and addresses shortfalls with three elements:

- **Context**: involves determining the scope of mission success, as well as negative effects, during operations. This element includes operational trends, detailed analysis on specific aspects of operations, and a diagnostic "report card" for senior leaders on progress made in the two PPG objectives discussed above.

- **Cause**: involves identifying specific factors (root causes) that lead to undesired effects to facilitate learning lessons from the past and inform adaptation.

- **Conduct**: marries *cause* with *context* to identify tailored changes to guidance and approach to improve the success of future missions while minimizing unintended consequences.

The first element provides context and directly addresses the desired balance in the PPG. It answers such questions as:

Figure 5. Analytical approach to improve lethal force operations

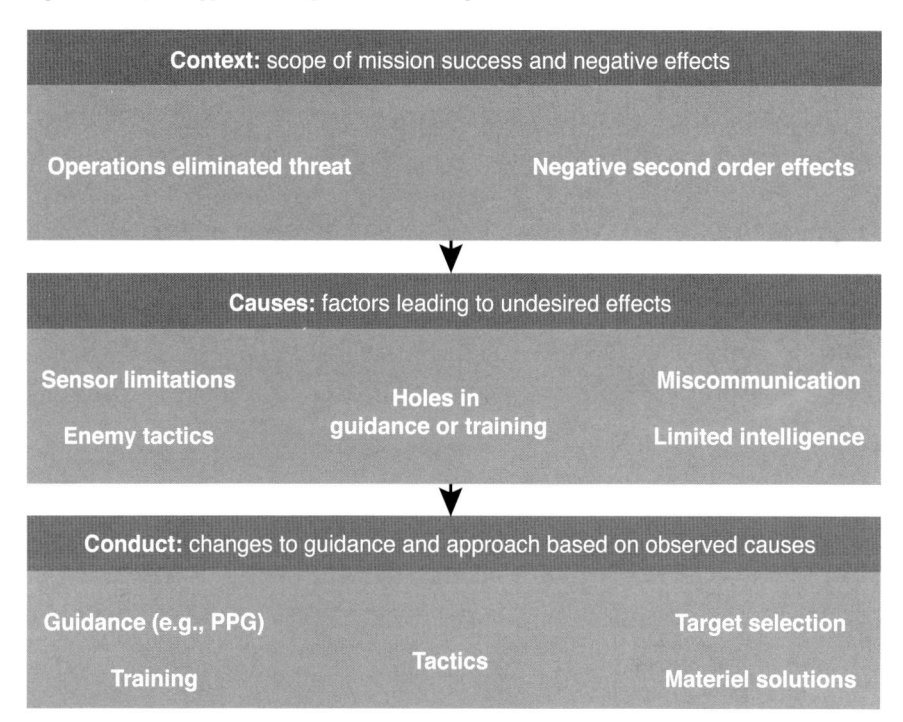

Context: scope of mission success and negative effects

Operations eliminated threat — Negative second order effects

Causes: factors leading to undesired effects

Sensor limitations — Holes in guidance or training — Miscommunication
Enemy tactics — Limited intelligence

Conduct: changes to guidance and approach based on observed causes

Guidance (e.g., PPG) — Tactics — Target selection
Training — Materiel solutions

- How effective are operations overall?
- How do operations by type and action arm vary in mission success?
- How frequent are civilian casualties during operations?
- When do they tend to occur?

The second element answers the question of "why," for example:

- Why did the missions not successfully eliminate the intended target?
- When operations cause civilian casualties, what are the contributing factors that played a role?

Finally, by comparing the first two factors, we can arrive at an understanding of specific areas for improvement to inform subsequent guidance and operational approaches—and not repeat mistakes.

As will be shown, this approach offers a number of positive outcomes. It can ensure guidance is best suited to meet its intent and inform revisions to that guidance over time. It can also achieve the traditional outcome of operations research, which is to improve the operational effectiveness of U.S. lethal action missions. The following sections examine the components of the framework in more detail, and illustrate them with data. Since information on these operations is classified, the analysis here cannot use official operational data. Where open source data are available, however, they will be used to illustrate the analytical approach.

ELEMENT 1: CONTEXT

For the first element, operations are examined to quantify the scope of both mission success and negative impacts that occur during operations to understand current performance and identify potential areas where improvement is possible. This analysis includes: mission success; civilian casualties; and a diagnostic "report card" for senior leaders on progress made regarding the two PPG objectives.

Mission Success

The first step in establishing context is to examine mission success. For our purposes, that means characterizing lethal action operations and analyzing their operational effectiveness. There are three components to this characterization. The first is mission profiles, or the number and nature of operations. Mission profiles include a description of:

- operational tempo: the number of operations over time;
- types of targets: both physical (e.g., vehicles, buildings) and organizational (e.g., senior leaders, facilitators);

- types of operations conducted: manned airstrike, drone strike, raid, etc.; and

- subsets of the above for operations conducted by specific forces, military or otherwise.

The second component is mission effectiveness, which can be viewed as progress toward the positive impacts of lethal action shown in table 4. For example, for the goal of killing a particular individual who has been identified as a threat, one appropriate measure could be the success rate for lethal action against targeted individuals.[1] Another measure is the number of combatants killed per strike. While attrition is not the primary goal of U.S. CT operations using lethal force, attrition can disrupt terrorist networks by removing trained militants from the battlefield and causing a deterrent effect to prospective recruits, so tracking this measure can be useful.

The final component to be assessed is the number of proposed targets for which an action (e.g., raid or airstrike) could not be completed. Two sets of targets are relevant here: actions that could not be taken because of insufficient intelligence for the strike, and those that could not be acted upon because of PPG restrictions, such as civilian casualties. This distinction between targets helps characterize the magnitude of the barrier to conducting operations due to intelligence requirements and other PPG restrictions. The analysis should also include a discussion of adversary approaches that complicate lethal action and targeting criteria. Determining these adversary countermeasures overall and for specific periods—either annually or (optimally) according to dates that delineate changes in policy and guidance—helps understand the interplay between these changes and overall mission success.

These measures should be evaluated for the total set of operations as well as for key subsets, including:

- specific target types (e.g., building, vehicle, gathering);

---

[1] Estimates of progress on other short- and long-term goals will necessarily rely on intelligence to estimate the wider effects of lethal action.

- specific echelons of targets (e.g., leaders, mid-level commanders, specialists, etc.);

- nature of operations (e.g., drone strike, manned aircraft strike, unilateral raid, partnered raid, combination);

- preplanned versus fleeting targets; and

- comparing operations by different action arms.

This analysis will highlight areas of success, as well as point out potential areas for improvement. For example, examining the subsets listed above may provide additional information on factors that drive overall trends (e.g., specific types of targets that tend to impact mission success or potential differences between preplanned and fleeting target sets). This examination both informs the conduct of future operations and focuses root cause analysis (discussed later) to determine specific reasons for these differences. In addition, such analysis informs operational decision making for prospective operations so that historical factors are accurately taken into account.

These measures provide insight into near-term targeting performance, particularly in terms of the country's ability to successfully take specific targets off the battlefield. There is also a need to assess the effect of these operations on the threats to the nation that predicated the lethal action, which requires a more intelligence-centric analytical effort that assesses threat streams and the likely effects of past U.S. operations. This effort should include examples of specific terrorism plots disrupted through lethal operations, as well as an assessment of the broader effect of lethal force on enemy networks.

Civilian Casualties
The other major objective of the PPG focuses on ensuring that noncombatant civilians are not killed or injured during U.S. lethal action operations. Assessing civilian casualties during lethal action operations begins with a

raw total number of civilian casualties, which can be compared across different periods (years, or possibly other dates that coincide with changes in guidance or operational approaches). While civilian casualties technically include both killed and wounded, typically it is more difficult to obtain totals of wounded than it is for those killed.[2] So, in cases where wounded estimates are difficult to obtain reliably, the number of civilians killed usually acts as a surrogate metric for total civilian casualties.[3]

Although the total number of civilian casualties is useful for understanding the magnitude of the toll of lethal actions on civilians, it is insufficient for determining progress. For example, ISAF tracked the total number of civilian casualties from its operations per year and touted progress when the number decreased, but this data neglected significant factors—such as operational tempo and operating environment—that can also affect the toll on civilians. For example, if civilian casualties decreased by 10 percent over some period but the number of operations during that time dropped by 50 percent, then the metric for total civilian casualties shows apparent progress. Yet in this case, an average operation would have been more likely to cause civilian casualties after the decrease in the total, putting into question the claim of improved care during operations. Normalizing the number of civilian casualties by the number of operations (or a rate of civilian casualties per operation) can put casualties in operational context and indicate whether operations are more or less likely on average to cause such casualties over time. Similarly, the rate of civilian casualty incidents per the total number of operations indicates the relative propensity of operations to cause civilian casualties.

Another measure focuses on the relative lethality of incidents, or how many casualties on average result from an operation where civilian casualties occur. For example, in Afghanistan, when civilian casualties resulted

---

[2] Reasons include varying definitions of injuries that count as wounded, as well as the greater chance of injuries going unreported.

[3] For more on this topic, particularly ISAF's role in measurement and the Civilian Casualty Tracking Cell, see Bob Dreyfuss and Nick Turse, "America's Afghan Victims," *Nation*, 18 September 2013, http://www.thenation.com/article/americas-afghan-victims/.

during checkpoint operations, the average number of casualties remained relatively small (e.g., approximately one casualty for incidents involving individuals, and two casualties for incidents involving vehicles). In contrast, airstrikes tended to result in a higher number of casualties per incident. In addition to highlighting differences in civilian casualties based on the types of operations, this measure was also useful in showing greater effectiveness in guidance over time in Afghanistan, as the average number of casualties per civilian casualty incident—an operation where civilian casualties occurred—dropped significantly following the issuance of improved guidance.[4] For context, another useful measure is the average number of civilian deaths per operation, which illustrates the average number of civilian deaths that could be expected to result from a typical operation.

These measures should also be broken out by type of operation and by action arm. Similar to the measures for mission success, this analysis will highlight areas of success as well as point out potential areas for improvement with respect to avoiding civilian harm during operations. This measure provides context to the subsequent stages of the analytic framework described later.

Report Card: Relationship between Mission Success and Civilian Casualties
The two objectives of improving mission success and reducing civilian casualties may be in conflict at times, and indeed it can be a considerable challenge to conduct operations effectively and simultaneously minimize such negative second-order effects as civilian casualties (figure 6).

In this scenario, an effort to increase mission success from condition A to condition B would carry a commensurate cost of an increase in civilian casualties. This scenario also implies that the requirement to reduce civilian casualties would induce a decrease in mission success.

Yet while this assumption has been voiced by some military forces directly, and by news reporting indirectly, the relationship between mission success and civilian casualty rates is not necessarily always direct. In fact,

---

[4] Lewis, *Civilian Casualties: Enduring Lessons*.

Figure 6. Perceived tension between mission success and CIVCAS

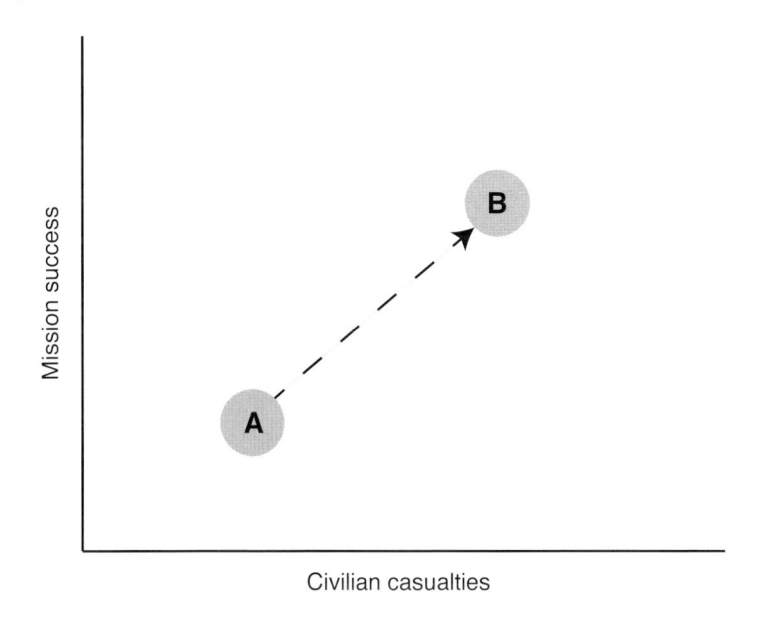

Figure 7. Common relationship between mission success and CIVCAS in Afghanistan

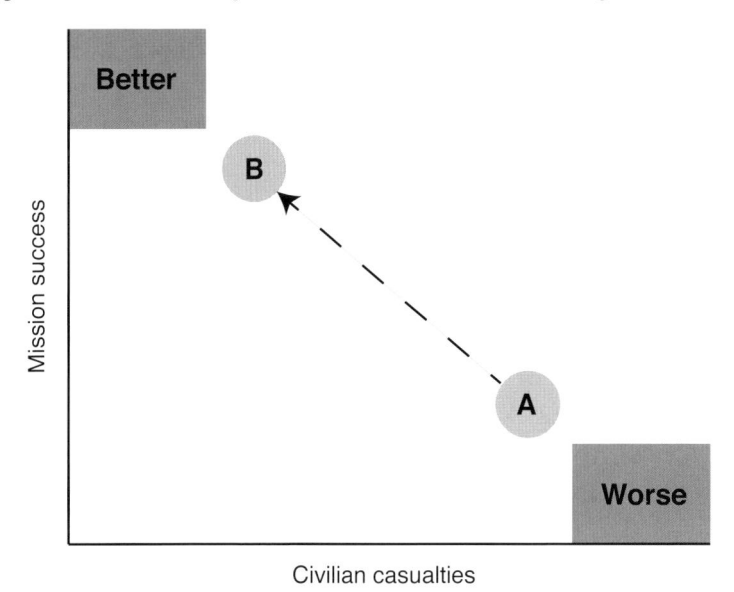

An Analytical Approach to Improve Lethal Action Operations | 121

for CT operations in Afghanistan, the relationship appeared to be more in line with that depicted in figure 7.

As illustrated above, an effort to improve mission success would also decrease civilian casualties, which would subsequently increase mission success. The optimum outcome is represented by the upper left quadrant of the chart. Fortunately, these two objectives can in some ways be mutually reinforcing. For example, many of the measures taken to reduce civilian casualties—improved pattern-of-life analysis, improved intelligence requirements, better coordination, and situational awareness—also work to enhance mission success. However, realizing that such a relationship exists came as a result of careful attention to improving mission success, reducing civilian casualties, and optimizing guidance and operational approaches to this end. Such a mutually reinforcing relationship between these two goals, where they both improve simultaneously, is unlikely to result without an intentional and informed effort.

Given the twofold criteria of the PPG aiming for both mission success and avoiding civilian casualties, it is worthwhile to determine the relationship between these two factors. A hypothetical example using notional data can be seen in figure 8.[5]

In figure 8, operations from 2010 to 2011 and from 2013 to 2014 move toward the upper left quadrant of the chart as mission success and avoidance of civilian casualties improve simultaneously, consistent with the aims of the PPG. From 2012 to 2013, however, an improvement in reducing civilian casualties came at the same time that mission success decreased; and from 2011 to 2012, decreases are apparent in both PPG goals.

These results are summarized in table 5; this kind of table can be used as a report card for senior leaders regarding progress in meeting the two PPG objectives. In the table, each year is color-coded to show the extent to which operations achieved the aims of the PPG per changes in operations that year compared to the previous year. Green indicates both an increase

---

[5] Note that, while the PPG is dated May 2013, the stated aims of the PPG—effectively dealing with the threat and avoiding civilian casualties—had been discussed for years prior to the release of this guidance.

Figure 8. Relationship between mission success and CIVCAS

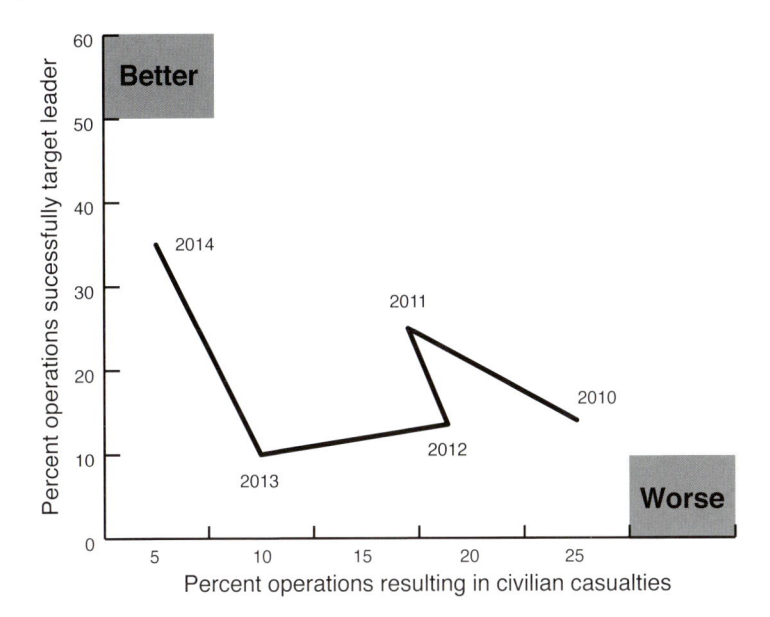

Table 5. Hypothetical report card for meeting PPG goals

Year	Mission success improved?	Civilian casualties less likely?
2011	✓	✓
2012	✗	✗
2013	✗	✓
2014	✓	✓

An Analytical Approach to Improve Lethal Action Operations | 123

in mission success (or no significant change) and a reduction (or no significant change) in the rate of civilian casualties; red indicates that mission success decreased while the rate of civilian casualties increased; and yellow indicates that one of the two PPG aims was achieved but not the other. Based on data shown in the figure above, the table shows that operations in 2011 and 2014 were more successful in approaching the aims of the PPG, while there was room for progress in 2013. In contrast, operations in 2012 met neither of the PPG aims.

Note that there may be good reasons for a period of operations coded yellow or even red. For example, a change in enemy tactics to hide within the population could both increase the rate of civilian casualties and increase the number of cases where civilians were misidentified as terrorists, resulting in less successful operations. Consequently, the context step should be accompanied by the next step—cause—described below.

While this approach uses changes from year to year to assign colors, it is also possible to assign absolute requirements for operations and then grade operations relative to those requirements. For example, operations could be assigned threshold requirements of 70 percent for mission success and 1 percent for a civilian casualty rate. In this arrangement, the operational performance would be graded according to those stated requirements. This graded approach has the advantage of avoiding cases where excellent performance one year does not increase the likelihood of a marginal or failing grade the next year. However, such thresholds would require up front agreement. If this approach were followed, reasonable thresholds should be determined through a process referencing previous operational data. For example, civilian casualty rates were determined for several kinds of operations in Afghanistan, so such data already exists and should be consulted.

ELEMENT 2: CAUSE

This element of the analytic approach involves identifying specific factors—root causes—leading to undesired effects to facilitate learning lessons from the past and to inform adaptation. The previous stage—

context—highlights areas of progress and areas where improvement is possible. However, further analysis is needed to identify the root causes of areas where operations fell short of goals, which can then be used to develop tailored improvements to guidance and operational approaches (i.e., the conduct element).

This process begins with data-centric reconstructions of individual incidents of interest: incidents where mission success was not achieved, civilian casualties were incurred, or both. Operational data should be a primary source, but data from other sources should also be factored in, including that from international organizations (e.g., the UN, ICRC), nongovernmental organizations, and media reports. After the reconstruction step, which results in identifying contributing causal factors, further analysis finds common trends in these causal factors to ascertain the main root causes of specific outcomes.

The process described here is distinctly different than that for a standard after action report (AAR) process, where any root cause analysis is done for a single incident in isolation, and lessons are typically not effectively shared across different units or operations. In contrast, by collectively considering causal factors for multiple incidents by different units and operations, as is done here, this process can reveal larger patterns to focus efforts for reducing civilian harm or improving mission success on the most productive areas. In Afghanistan, that collective analysis process revealed a number of factors leading to civilian casualties that had not been previously identified, such as certain target sets not being addressed in guidance, certain scenarios not being adequately covered in doctrine, and equipment/capabilities not supporting the targeting process for select situations.

ELEMENT 3: CONDUCT

This element looks at ways to change the conduct of future operations to promote both mission success and reduced civilian casualties through tailored modifications to guidance and operational approaches. It should be noted that this aim is partially achieved through regular adaptation that occurs with a learning force. For example, in Sadr City, Baghdad, Iraq, in 2008, U.S. forces combined new tactics, massed and fused intelligence, precision weapons, and other innovative procedures and command and control arrangements to mitigate civilian casualty concerns while also aiding the tracking of targets. This adaptation provided freedom of action to use force effectively against a fleeting enemy operating in an urban area.[6]

At the same time, many examples show where adaptation came slowly or inconsistently within the force. Thus, the analytic framework and lessons learned process described here may accelerate and optimize the normal learning process for operations. This analysis recommends tailored changes to guidance and approach based on actual observed causes, taking the root causes identified above and informing them with operational trends, pointing out specific areas of challenges and their relative frequency and importance. These changes represent concrete steps to address those challenges in future operations. One such challenge in guidance is incorporating specific scenarios that were not adequately addressed, putting in place tactics or procedures to improve the fidelity of intelligence, deciding on a specific action arm or weapon platform, improving training for a certain kind of operation, or elevating risk factors associated with specific target sets or environments.

Importantly, changing conduct to adapt to new lessons learned should not be a one-time process. As causes are successfully identified and addressed in operations, improvement should result. Yet new factors also may arise as tactics change or the enemy adapts. Hence, the process should be repeated periodically—say, every 6 or 12 months—where performance is reassessed, the need for additional changes is identified, and guidance and operational approaches are adapted accordingly.

---

[6] Lewis, *Civilian Casualties: Enduring Lessons*.

CHAPTER 15

ILLUSTRATING THE APPROACH

Previous chapters of Part III described the analytic framework in conceptual and theoretical terms. This chapter takes the theory and puts it into practice using available data to illustrate the framework and lessons learned approach and its potential benefits. Ideally, data for a single case study would be available to demonstrate the approach from start to finish. However, such is not the case (at least publically) for U.S. lethal-action CT operations. The specific examples we use to demonstrate the approach are as follows:

- **Context:** analysis of mission success and civilian casualties using open-source data on U.S. operations in Yemen and Pakistan, followed by a report card for those operations.

- **Cause:** a vignette of a high-profile civilian casualty incident, followed by root-cause analysis of civilian casualties from drone operations in Afghanistan.

- **Conduct:** a discussion of specific changes to guidance and operational approaches made in Afghanistan that were prompted by analysis.

These examples aim to show how this process can yield additional insight into operational performance and enable improved adaptation and learning to promote mission success and reduced civilian casualties in future operations.

CONTEXT: YEMEN AND PAKISTAN

One salient aspect of assessing the context of operations is the mission profile. Figure 9 shows the number of U.S. counterterrorism operations per year in Pakistan and Yemen, using available data maintained by New America Foundation (NAF).[1] These data provide information about operational tempo, which is useful for understanding the ability to conduct operations—depends on intelligence and freedom of action—as well as the overall context for more detailed analysis. Note that the figure shows data for the first half of 2014, so the significant drop in 2014 compared to 2013 is at least partially due to the shorter time period. In addition, no operations took place in Pakistan for the first six months of 2014.

From the figure, it is clear that the number of overall operations has decreased in the past few years. From 2010 to 2012, an average of 100 operations took place per year. During this time, the operations shifted from being almost entirely in Pakistan to being divided equally between Pakistan and Yemen. In 2013, however, the number of operations decreased by about half of previous levels; operations in 2014 were on pace for further reduction.

Not all data elements mentioned in the previous chapter—such as types of operations and the identity of units involved—are available in open-source data. However, target types are included in data extracted from media reporting, so analysis of that element will be used here as an example for the overall analytic process. Operations in Pakistan and Yemen featured a number of different target types: buildings (homes, businesses, and madrassas), vehicles (cars and motorcycles), a combination of vehicles and buildings, and gatherings of people. Figure 10 shows the distribution of target types for operations in Yemen and Pakistan, as characterized in

---

[1] See Singh, "A Meta-Study of Drone Strike Casualties."

Figure 9. Lethal action operations per year in Pakistan and Yemen

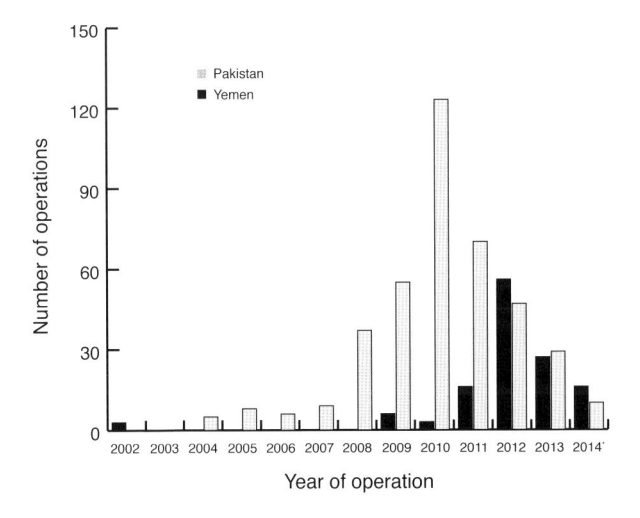

Source: Data from New America Foundation as of 4 August 2014 on http://securitydata.newamerica.net/.

Figure 10. Target type distribution for lethal action operations in Pakistan and Yemen, 2012–14

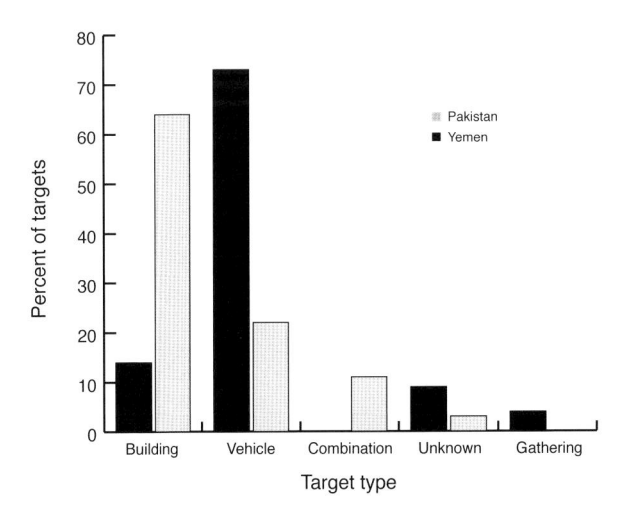

Source: Data from New America Foundation, http://securitydata.newamerica.net/.

Illustrating the Approach | 129

NAF's database.[2] Some target types could not be determined, which are included in the figure as "unknown."

Per figure 10, target sets for operations in Yemen and Pakistan are significantly different—most of the operations in Pakistan targeted buildings, while operations in Yemen targeted vehicles. In addition, about 1 in 10 operations in Pakistan targeted a combination of buildings and vehicles simultaneously; no such events occurred in Yemen over the timeframe analyzed.[3] Similarly, several events in Yemen targeted gatherings of people, with at least one of them believed to be a terrorist training camp.

Another component of assessing the context of operations is mission success, which has a number of different aspects. This section focuses on the relationship between mission success and target types to illustrate the larger analysis process with available data. For example, figure 11 shows the success in killing senior leaders by target type (when known) for operations in Yemen and Pakistan.[4]

Overall, operations in Yemen were more likely to successfully target senior leaders. In addition, actions against different target sets were more successful in Yemen and Pakistan. For Yemen, an operation targeting vehicles was about three-and-a-half times more likely to result in the death of a senior leader than an operation targeting buildings. This statistic is quite different from Pakistan, where targeting success for senior leaders was about equal for both target types.

Another metric to consider is the total number of terrorists (legally, characterized as combatants) killed during an operation. Figure 12 shows the average level of attrition from operations in Yemen and Pakistan, both

---

[2] These figures are based on operations in Pakistan and Yemen between 2012 and 2014 to highlight any country-specific differences, since there were considerable numbers of operations occurring in both countries during that timeframe. Data on individual operations came from NAF.

[3] Only one combination event took place during U.S. operations in Yemen, occurring on 14 October 2011. This event led to the death of two senior leaders and also resulted in an estimated 9–15 civilian casualties.

[4] This aggregates NAF data for individual airstrikes, indicating target type and whether senior leaders were killed during individual strikes. Determination of "senior leader" was made in the NAF database.

Figure 11. Percent of lethal action operations that successfully targeted a senior leader, by target type

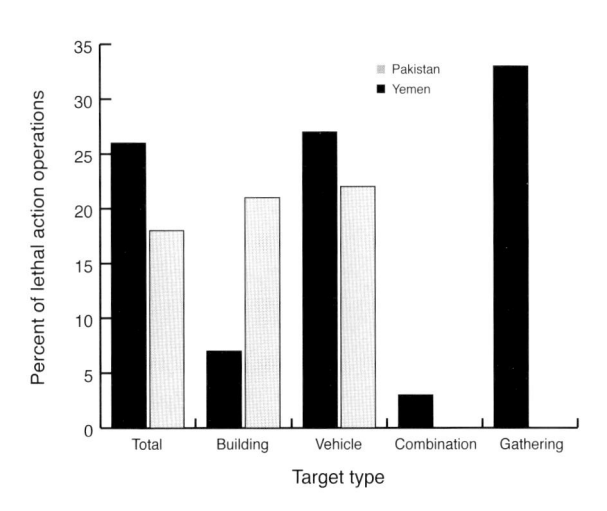

Source: Data from New America Foundation, http://securitydata.newamerica.net/.

Figure 12. Maximum/minimum combatant casualties from lethal action operations with different target types

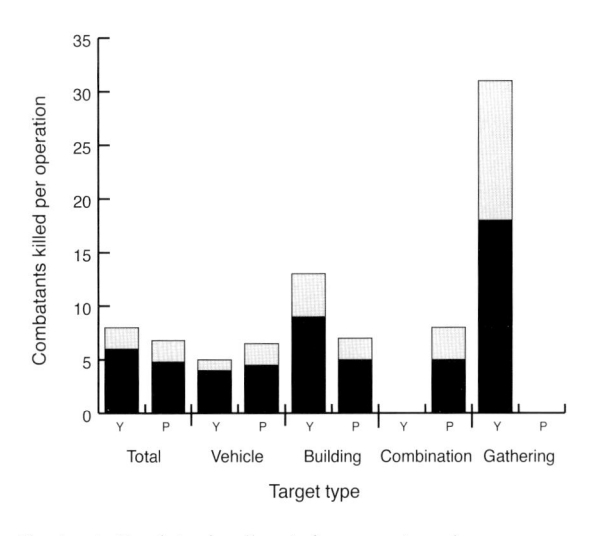

Source: Data from New America Foundation, http://securitydata.newamerica.net/.

Illustrating the Approach | 131

overall and for specific target types. The data source provides a minimum and maximum combatant casualty number for each operation based on differing media reports, so average minimum and maximum values are provided.

In figure 12, the average combatant numbers of those killed in action (KIA) are about the same for operations in Yemen and Pakistan. For different target types, the casualties from engagements at buildings tend to be larger than those for vehicles. Also, engagements at buildings in Yemen cause about twice as many casualties as those in Pakistan. This could reflect differing pattern-of-life factors—that combatants are more likely to congregate in buildings in Yemen vice Pakistan—or that weapons employment, either in terms of ordinance or tactics, is significantly different in the two operations. For example, if engagements in Yemen routinely employ 500-pound bombs, while engagements in Pakistan use Hellfire missiles, then a larger attrition effect could be anticipated in Yemen operations due to the larger area of weapon effect (all other things being equal). Combination events, such as an operation that targeted both a building and a vehicle, occurred primarily in Pakistan, and had casualty numbers close to that for building targets. Operations targeting gatherings showed casualty levels several times higher than those for other target types.

The discussion here focuses on specific types of targets and one echelon of targets (senior leaders). With access to official data, additional analysis could address the different echelons of targets and the relative success of preplanned versus fleeting targets. In addition, analysis could be done with a focus on the nature of the operation (e.g., drone strike, manned aircraft strike, unilateral raid, partnered raid, combination, etc.) and the relative success of these approaches. In addition, operations by different action arms could be compared in terms of overall mission success—the average percentage of operations that successfully dealt with their intended target—as well as in terms of specific kinds of operations to show strengths and possible areas of improvement for each force. Examining these subsets could provide additional information on factors that may drive overall trends (e.g., specific types of targets that tend to be less successful, or potential differences between preplanned and fleeting target sets). This both informs the conduct of future operations and focuses root cause analysis (discussed later) to determine specific reasons for these differences.

As mentioned above, another important component to be assessed is the number of proposed targets that could not be attacked, including those that could not be developed sufficiently to provide supporting intelligence for the strike and those that could not be acted upon because of other PPG restrictions. This analysis helps characterize the magnitude of the barrier to conducting operations due to intelligence requirements and other PPG restrictions. It should also include a discussion of adversary approaches that complicate lethal action and targeting criteria. Unfortunately, we cannot present sample analysis here for these elements due to a lack of open-source data.

The previous discussion addressed mission success, one of the major objectives of the PPG. The other major objective is preventing civilian casualties during U.S. lethal action operations. While U.S. official data on civilian casualties during these operations has not been released, open-source data are available and will be used here to illustrate the proposed analytical process for evaluating civilian casualties from the use of lethal force. Several sources are available for the drone strikes subset of U.S. CT operations, including the Bureau of Investigative Journalism (BIJ) and the New America Foundation (NAF). These two sources support agreement overall, particularly after treating the NAF "unknown" status casualties as civilian, per international humanitarian law considerations.[5] BIJ data are used here because they include operation types other than drone strikes, giving a more comprehensive dataset for the present analysis.

For illustrative purposes, these measures are included in table 6 using BIJ data for Pakistan and Yemen. These numbers show the peril of relying on only the numbers of civilian casualties—or in this case, civilian deaths—without other metrics to provide context. From the data below, civilian deaths in Pakistan from U.S. operations were more than twice those com-

---

[5] See Lewis, *Drone Strikes in Pakistan*, for more discussion of this point. The author confirmed the validity of this interpretation of IHL per a communication with ICRC, May 2014.

Table 6. Overall civilian death statistics for Pakistan and Yemen through August 2014*

	Pakistan	Yemen
Overall lethal action operations	384	176
Civilians killed (CIV K)	416	152
Operations where CIV were killed (CIV K incidents)	73	27
Average CIV K/operation	1.1	0.9
CIV K incidents/operation (percent)	19	15
CIV K/CIV K incident	5.6	5.6

* BIJ includes a range of values for casualty numbers. This table uses the minimum values to provide a conservative measure of the civilian toll.
Note: after 16 July 2014 through the end of August, there were seven drone strikes in Pakistan. For completeness, these operations were also included in the dataset analyzed here.
Source: Bureau of Investigative Journalism, current as of 16 July 2014 on https://www.thebureauinvestigates.com/category/projects/drones/drones-graphs/.

pared to deaths from operations in Yemen. When considering the other measures, however, operations in Yemen were only somewhat lower in their likelihood of causing civilian deaths (15 percent versus 19 percent) and caused the same number of deaths per incident.

These metrics can also be considered over time. For example, the percentage of total operations resulting in civilian deaths (CIV K incidents/operation from table 6) is shown in figure 13 for both Pakistan and Yemen. This metric is shown beginning in 2009 (the first year of CT operations in Yemen) through the first half of 2014. In this figure, the percentage of operations in Pakistan causing civilian deaths decreased over time, reaching zero in 2013 and 2014; the decline of civilian casualties is clearly within the intent of the PPG. In contrast, the likelihood of civilian casualties in Yemen operations varies over time, first dropping after 2009 from a maximum of 34 percent, but then ranging between 10 percent and 24 percent in subsequent years, in contrast to the clear drop over time in Pakistan. In fact, the civilian casualty rate in Yemen rises from a minimum in 2012 to double that rate in 2014, conflicting with the intent of the 2013 PPG.

Figure 14 shows the lethality of incidents (i.e., the average number of civilians killed when civilian casualties occurred). This metric is shown beginning in 2009 (the first year of CT operations in Yemen) through August

Figure 13. Percent of lethal action operations causing civilian deaths, by year

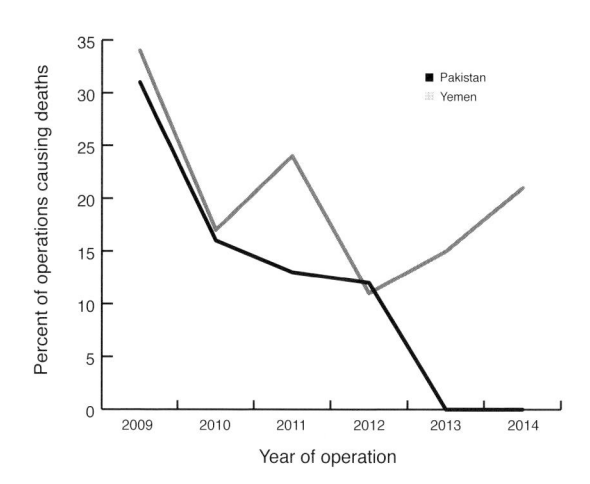

Source: Bureau of Investigative Journalism, https://www.thebureauinvestigates.com/category/projects/drones/drones-graphs/.

Figure 14. Average civilian deaths per incident, by year

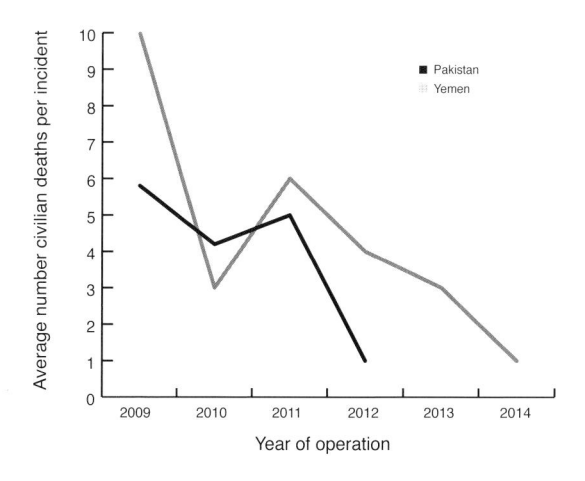

Source: Bureau of Investigative Journalism, https://www.thebureauinvestigates.com/category/projects/drones/drones-graphs/.

Illustrating the Approach | 135

2014. In this figure, five civilians were killed on average for an incident in Pakistan between 2009 and 2011; after this, the lethality of operations decreased sharply to one civilian killed on average in 2012 and zero in 2013 and 2014. For Yemen, the average in 2009 is very high due to a single incident with 44 civilian deaths. After this, the average lethality decreased to a range between three and six civilians killed per incident for the four-year period from 2010 to 2013, and then dropped further to an average of one-and-a-half killed per incident for the first half of 2014. Overall, both Pakistan and Yemen operations showed a decreasing trend in lethality over time, consistent with the intent of the PPG. For Yemen, however, the decrease in lethality was countered by the increase in the rate of civilian casualty incidents for operations in Yemen in 2014.

Figure 15 illustrates the average number of civilian deaths that could be expected to result from a typical operation. For Pakistan operations, the average civilian toll decreases steadily from 2009 down to zero in 2013 and 2014. For Yemen, after 2009 with its single high-casualty incident, this number ranges between 0.3 (in the first half of 2014) and 0.5 deaths per incident, with the exception of 2011, where it spiked to 1.5. This spike is due to a simultaneous increase in the civilian death rate and the lethality of incidents for that year. The lowest value for the first half of 2014 is driven

Figure 15. Average civilian deaths per operation

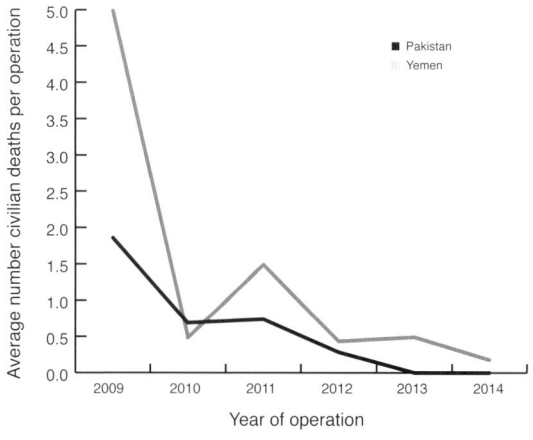

Source: Bureau of Investigative Journalism, https://www.thebureauinvestigates.com/category/projects/drones/drones-graphs/.

by the low lethality rate of incidents, despite an increase in the percentage of operations causing civilian deaths.

Overall, an analysis of open-source data for civilian deaths in lethal action operations in Pakistan and Yemen shows some progress in meeting the stated U.S. intent to reduce civilian casualties during operations. In Pakistan, the likelihood of civilian casualties per airstrike decreased over time, and dropped to zero after the release of the 2013 PPG. In comparison, the use of lethal force in Yemen has not shown the same reduction in civilian casualties; while the lethality of incidents decreased in the first half of 2014, the frequency of operations causing civilian casualties showed an increase in recent years, even after the issuance of the PPG.

The measures provided above should also be broken out by type of operation and by action arm. However, these statistics are generally not available in open-source data and cannot be illustrated here. But one area can be examined in more detail—characteristics of drone strikes versus other kinds of operations. BIJ is particularly interested in drone strikes; so for operations in Yemen, they have specifically researched CT operations to determine whether each operation represents a drone strike.[6] In some cases, they have reliable information that indicates an operation was in fact a drone strike. For other operations, BIJ cannot identify whether they are drone strikes, manned airstrikes, ground raids, or a combination thereof; for our purposes, these are referred to as "other operations." While it would be preferable to know the specific type of engagement for all operations, the existence of two groups—drone strikes and other operations—allows a comparison of civilian casualty attributes for one group (confirmed drone strikes) compared with another (unknown—possibly manned aircraft strike, drone strike, or ground operation) (table 7).

In Yemen, per this dataset, approximately 40 percent of operations fall under the category of "confirmed drone strikes." However, this type of operation is responsible for the majority of civilian casualties—more than 80 percent of the total number of civilian deaths. The table also provides some

---

[6] In Pakistan, all operations are believed to be drone strikes.

Table 7. Comparing CIVCAS from drone strikes and other operations in Yemen

Yemen	Confirmed drone strikes	Other operations
Overall number of operations	69	107
Civilians killed (CIV K)	124	28
Operations where CIV were killed (CIV K incidents)	19	8
CIV K/CIV K incident	6.5	3.5
Rate of civilian deaths per operation (percent)	27.5	7.5

Source: Bureau of Investigative Journalism, https://www.thebureauinvestigates.com/category/projects/drones/drones-graphs/.

reasons for this statistic: the group of confirmed drone strikes has almost twice the civilian deaths per incident compared to the other group; and drone strikes are almost four times more likely to cause civilian casualties per operation. While this observation should be considered tentative until the comparison is repeated with operational data, a similar trend—with a higher civilian casualty rate for drone strikes compared to other types of operations—has been seen previously. This finding will be discussed further in the benefits section.

Report Card for Operations in Yemen and Pakistan
To build a report card regarding operations in Yemen and Pakistan and to demonstrate how well they met U.S. objectives outlined in the PPG, the relationship between mission success and avoiding civilian casualties must first be determined (figure 16).

In figure 16, mission success is represented by the percentage of operations successfully targeting senior leaders. For Pakistan from 2012 to 2013, operations move toward the upper left quadrant of the chart, with mission success and avoidance of civilian casualties improving simultaneously, consistent with the aims of the PPG.[7] At other times, however, an improve-

---

[7] It could be considered unfair to find that, for example, operations in Pakistan in 2012 fell short of the full aims of the PPG, since the PPG did not exist then. However, it is useful in this context to consider historical performance and whether these aims were achieved during operations. At the same time, expectations should be higher for operations in 2013 and 2014, since these aims were more explicit in the guidance.

Figure 16. Relationship between mission success and CIVCAS

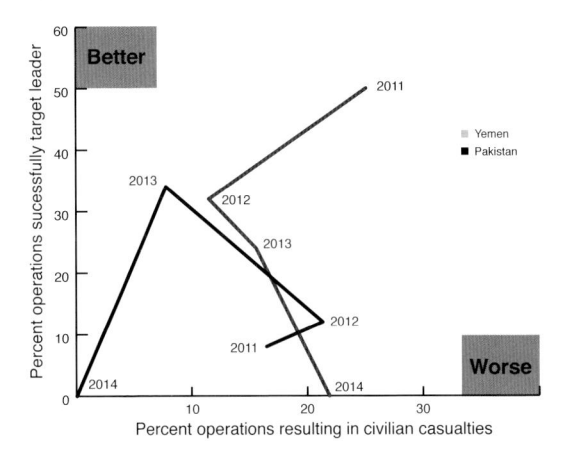

[Figure: Line chart plotting "Percent operations successfully target leader" (y-axis, 0–60) against "Percent operations resulting in civilian casualties" (x-axis, 0–30) for Yemen and Pakistan from 2011 to 2014. "Better" is indicated in the upper-left and "Worse" in the lower-right.]

Source: author analysis based on Bureau of Investigative Journalism data, https://www.thebureauinvestigates.com/category/projects/drones/drones-graphs/.

ment in one of the two criteria came at the expense of the other. For Yemen, a decrease in the frequency of civilian casualties from 2011 to 2012 came at the cost of decreased mission success. And, contrary to the aims of the PPG, from 2012 to 2014, mission success dropped over time while the civilian casualty rate increased. Thus, in Yemen, operations became less likely to satisfy both PPG criteria during this time.

These results are summarized in table 8, which shows year-to-year changes that are color-coded to illustrate the extent to which operations achieved the aims of the PPG. Green indicates both an increase in mission success and a reduction in the rate of civilian casualties; red indicates that mission success decreased while the rate of civilian casualties increased; and yellow indicates that one of the two PPG aims was achieved but not the other. Based on the table, Pakistan operations have been more successful in approaching the aims of the PPG, though there is still room for improvement in 2014. In contrast, Yemen operations over the past two years appear to be meeting neither of the PPG aims. From this report card, the analytic process described here could reasonably start with a focused effort supporting Yemen operations.

Table 8. Operations in Yemen and Pakistan achieve the aims of the PPG

	Yemen		Pakistan	
	Percent change in mission success*	Percent change in rate of CIVCAS	Percent change in mission success	Percent change in rate of CIVCAS
2012	-18	-14	+5	+4
2013	-9	+5	+21	-13
2014	-23	+5	-33	-7

*Measured here as the rate of success targeting senior leaders. With actual operational data, other metrics for mission success could also be used.
Source: author analysis based on Bureau of Investigative Journalism data, https://www.thebureauinvestigates.com/category/projects/drones/drones-graphs/.

CAUSE: EXAMPLES FROM AFGHANISTAN

To illustrate the process of root cause analysis, a specific vignette from Afghanistan is featured below, followed by an example of aggregating root causes from multiple incidents.

Uruzgan: Reconstructing an Incident of Interest

The vignette used here, which occurred in Uruzgan Province, Afghanistan, was chosen because it is well documented with sources approved for public release.[8] While the specific root causes identified below are not completely representative of CT operations in Yemen and Pakistan, the process for identifying such root causes is the same, so the process of reconstruc-

---

[8] Primary sources for the section on the Uruzgan incident come from the redacted CENTCOM report (*AR 15-6 Investigation*, 21 February 2010, CIVCAS incident in the vicinity of Shahidi Hassas, Uruzgan Province, Kabul, Afghanistan: 21 May 2010) and the unclassified ISAF executive summary ("Executive Summary" for *AR 15-6 Investigation*, 21 February 2010, CIVCAS Incident in Uruzgan Province, Kabul, Afghanistan: May 2010), hereafter *AR 15-6 Investigation*. For more information, see majGen Mchale, "Memorandum for Commander, United States Forces-Afghanistan/International Security Assistance Force, Afghanistan" (Kabul, Afghanistan: 2010), http://www.rs.nato.int/images/stories/File/April2010-Dari/May2010Revised/Uruzgan%20investigation%20findings.pdf.

tion and identification of contributing factors is illustrative of what could be done for those operations.[9]

Early on 21 February 2010, a U.S. Special Forces team, accompanied by Afghan Army and Police personnel, conducted an air infiltration into western Uruzgan Province for a daytime cordon-and-search operation. The Special Forces team was supported by a General Atomics MQ-1 Predator UAV and a Lockheed AC-130 gunship. While waiting for daylight, the team received intelligence that enemy forces were going to attack. The Predator operator observed two SUVs driving south toward U.S. forces and assumed these were enemy combatants. The position of the U.S. team and the initial position of the convoy are shown in figure 17.

The Predator showed the SUVs as they drove around the area for about three-and-a-half hours, changing directions several times, stopping to allow the occupants of the vehicles to pray, and moving to a position about 12 kilometers away from U.S. forces. The two SUVs were then joined by a third vehicle, a pickup truck. During this time, imagery analysts examined the full-motion video from the Predator and provided their feedback to the Predator crew. In turn, the Predator crew communicated with the U.S. team on the ground in Afghanistan.[10] The movement of the convoy is shown in figure 18.

Over three-and-a-half hours, the imagery analysts communicated the presence of three possible weapons in the vehicles and two children, also described as "adolescents." The descriptions provided by the imagery analysts—who were trained to interpret the Predator feed—were frequently different than the descriptions the Predator crew provided to the U.S. forces on the ground in Afghanistan. For example:

---

[9] While root causes are not identical for the two sets of operations, considerable overlap exists, so historical work supporting U.S. efforts in Afghanistan and products derived from it can be a useful starting point for the effort described here. For example, the checklist for AAR data elements in *Civilian Casualty Mitigation*, ATTP 3-37.31 (Washington, DC: Headquarters, Department of the Army, 2012), Appendix B, was derived from this Afghanistan work.

[10] Note that the imagery analysts were at a different location than the Predator crew, and had no means of direct communication with forces on the ground. Thus, their feedback was filtered through the Predator crew.

- Imagery analysts reported seeing several children, who were described as "adolescents." The Predator crew described them as "teenagers" who could be providing support to combatants, and later described them as "military-aged males."

- While the imagery analysts reported that the vehicles appeared to be leaving the area, the Predator crew stated that they might be flanking the U.S. position.

- At one point, the Predator crew communicated to the U.S. Special Forces team that the imagery analysts had reported seeing two weapons in a specific vehicle, when in fact no such report had been provided.

The imagery analyst descriptions were provided as text on computers via mIRC (Internet relay chat), which was not accessible to the Special Forces team, though it was available at their higher headquarters in Kandahar. The team then received additional intelligence indicating the enemy may be setting up an ambush. As a result, the Special Forces team believed the vehicles represented an imminent threat and called for an air strike (figure 19).

Bell OH-58 Kiowa Warrior helicopters were called in to strike the targets based on information provided by the Predator crew, including the presence of weapons and military-age males, but no mention of children. The vehicles were engaged with Hellfire missiles, with follow-up engagement of individuals using rockets. The OH-58 pilots then saw people running from the vehicles dressed in brightly colored clothing, which is characteristic of women's apparel in Afghanistan. So, they stopped the engagement and radioed back the possibility of civilian casualties. This communication was heard at the higher headquarters, but it did not include a report to ISAF with the possibility of civilian casualties as they were waiting to receive confirmation. This delayed the reporting of the incident for many hours, slowing the eventual ISAF consequence management response and causing that response to trail Taliban information operations

Figure 17. Initial locations of U.S. forces and civilian convoy

Source: Map data from Google TerraMetrics; information on locations based on redacted CENTCOM *AR 15-6 Investigation*.

Figure 18. Movement of civilian convoy

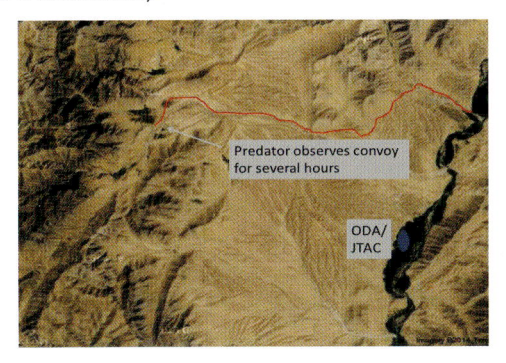

Source: Map data from Google TerraMetrics; information on locations based on redacted CENTCOM *AR 15-6 Investigation*.

Figure 19. Strike on civilian convoy

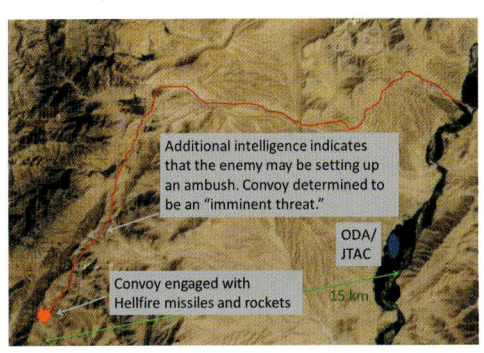

Source: Map data from Google TerraMetrics; information on locations based on redacted CENTCOM *AR 15-6 Investigation*.

and rumors rather than being proactive and establishing the facts quickly to avoid misunderstandings.

Lessons from the Uruzgan Incident
The "Swiss cheese model" of accident causation describes how problems (holes in the Swiss cheese) can arise relatively frequently with no impact; however, when the holes all line up, then an accident can occur. The Uruzgan incident, like other civilian casualty incidents, was the result of a number of factors that all contributed to the incident. These factors included:

- **Intended target not located due to lack of capability.** Specifically, the team on the ground received intelligence regarding the presence and intent of enemy forces but lacked the knowledge of where the enemy was located, which could have helped U.S. forces successfully deal with the threat and avoid misidentifying civilians as enemy combatants.

- **Misunderstanding the situation by different military elements.** The engagement was coordinated among the U.S. Special Forces team on the ground, its higher command element in Kandahar, the Predator crew at Creech Air Force Base in Indian Springs, Nevada, the Predator's support elements (including its imagery analysts at Hurlburt Field, Florida), and the OH-58 helicopters. Key facts were not shared among these different actors; as a result, the factors were not adequately considered in the decision to engage.

- **Predator crew overruled assessments by supporting imagery analysts.** The U.S. official investigation stated that the Predator crew, lacking training in interpreting imagery, "made or changed key assessments . . . that influenced the decision to destroy the convoy."[11] While the Predator crew was supported by imagery analysts who did have such training, only the Predator crew was in communication

---
[11] *AR 15-6 Investigation.*

with forces on the ground. Consequently, the descriptions they provided carried the weight of the entire processing, exploitation, and dissemination (PED) process that the Predator platform was supposed to provide.

- **"Guilt by association."** There were no facts to support declaring the third vehicle joining the convoy as hostile. Rather, the Predator crew described its status as "guilt by association." This assumption by the Predator crew, one that proved incorrect, was a violation of both the rules of engagement (ROE) and international humanitarian law (IHL).

- **Civilian casualty numbers and target types.** Until 2010, ISAF efforts to reduce civilian casualties from airstrikes emphasized particular target sets—namely, compounds—in its guidance and emphasis. The Uruzgan incident, along with others, such as the Kunduz tanker truck airstrike in September 2009, illustrates how significant numbers of civilian casualties could occur outside of compounds.[12] Consequently, the possibility of high numbers of civilian casualties was extended to other target types.

- **Lack of tactical patience.** While U.S. personnel watched the vehicles for three-and-a-half hours prior to engaging, which appears to show patience, the Predator crew appeared eager to engage and used leading language to describe the vehicles in terms that appeared to satisfy the ROE:

    - Responding to the imagery analyst description of seeing a child: "At least one child. . . . Really? Assisting the MAM [military-age male], that means he's guilty."

    - Responding to a later description of seeing several children: "I really doubt that 'children' call, man, I really (expletive deleted) hate that."

---

[12] For more on the tanker truck incident, see Stephen Farrell and Richard A. Oppel Jr., "NATO Airstrike Magnifies Divide on Afghan War," *New York Times*, 4 September 2009, http://www.nytimes.com/2009/09/05/world/asia/05afghan.html?pagewanted=all&_r=0.

- "That truck would make a beautiful target."

- "I want this pickup truck of dudes. . . . I hope we get to shoot the truck with all the dudes in it."[13]

- **Self-defense engagement predicated on an imminent—but not immediate—threat.** ROEs grant the authority to exercise self-defense in the face of a hostile act or hostile intent, which includes the threat of imminent use of force. This engagement was predicated on the interpretation of a single word: imminent. The definition of *imminent* in the U.S. military's standing ROE is not the commonly understood dictionary definition: threatening to occur immediately; near at hand; impending. Indeed, the standing ROE definition specifies that imminent need not be instantaneous or immediate. This distinction allows U.S. forces to take a broader view of what constitutes hostile intent and self-defense. In this incident, U.S. forces used this broad view to justify the use of force in self-defense, even though the vehicle convoy was miles and hours from Coalition forces. Hence, based on the guidance provided in the *AR 15-6 Investigation*, "there was no urgent need to engage the vehicles."

Collectively, these contributing factors represent areas where the U.S. military can make improvements. For example, training and doctrine can emphasize the importance of including key details in air-to-ground coordination. At the same time, training for drone communications could be revised. For example, investigate the inclusion of imagery analysts in voice communications so that the Predator crew is not forced to be the middleman in relaying intelligence they are not trained to interpret. Of course, these are simply potential solutions from a single incident; it is best to consider a larger set of incidents and then develop solutions that best fit the collective contributing causes. These solutions may include changes in

---

[13] *AR 15-6 Investigation.*

overall guidance (including the PPG), training, tactics, techniques, and procedures (TTP), and materiel solutions.

Aggregating Root Causes: Drone Strikes in Afghanistan
The other step in the root cause process aggregates individual incidents to identify common themes and primary contributing factors for operations that fail to meet their objectives. The previous analysis of drone strikes in Afghanistan operations offers a good example of this step.[14] As mentioned earlier, this analysis found the relative rate of civilian casualties to be 10 times higher for drone strikes compared to that for manned airstrikes (an example of the "context" stage of analysis). Then, all operations with civilian casualties were reconstructed in a way similar to the Uruzgan example given above. Finally, the contributing causes were aggregated to identify the main drivers of civilian casualties in drone strikes.

When multiple incidents were considered, several common root causes stood out that contributed to civilian casualties during drone strikes, including the following:

- **Training deficiencies for drone operators and imagery analysts.** A lack of sufficient training in patterns of life, positive identification, discriminating civilians from combatants, and tactical patience can increase the risk of civilian casualties.

- **Complex coordination processes.** The distributed PED architecture for drones commonly used to support decision making places higher demands on communications and coordination of engagements. Breakdowns in communication can lead to engagements that are not informed by the entire set of facts. Specifically, someone was aware of information that would have ended the engagement if it were commonly known.[15]

---

[14] Lewis, *Reducing and Mitigating Civilian Casualties*.

[15] This is also a common factor in friendly fire incidents, such as the shooting down of a U.S. Navy McDonnell Douglas F/A-18 Hornet jetfighter by a Patriot missile during Operation Iraqi Freedom in 2003. See Thomas E. Ricks, "Investigation Finds U.S. Missiles Downed Navy Jet," *Washington Post*, 11 December 2004.

- **Lack of situational awareness (SA) in the area beyond the immediate target.** Such a lack of SA was compounded by a well-intentioned tendency to use weapons that reduce collateral damage but did not always eliminate the target in the first engagement. A requirement for subsequent engagements increased the risk of civilian casualties because of unobserved civilians in the general area, often acting as first responders.[16]

This analysis informs subsequent efforts to tailor solutions to improve the conduct of operations, as described in the next section.

CONDUCT: EXAMPLES FROM AFGHANISTAN

These analytical steps were conducted for ISAF and U.S. troops in Afghanistan to help those forces reduce civilian casualties resulting from operations. For example, the Joint Civilian Casualty Study identified a shortcoming in the COMISAF tactical directives issued between 2007 and 2009.[17] The study findings were communicated to the new COMISAF, General David H. Petraeus, in mid-2010, and the revised tactical directive issued by his staff corrected the shortcoming, which had persisted through four previous versions of the guidance. As a result, the conduct of operations changed to reflect this key identified lesson.

By way of another example, in a 2011 study, analysis ascertained which kinds of operations contributed the most to civilian casualties, and what practical measures could be taken to reduce them.[18] After analyzing several hundred separate incidents, the study team provided a list of primary causal factors for different types of operations—including airstrikes, check point operations, artillery fire, and vehicle movements—and specific recommendations for changes in guidance and tactics to address them. ISAF made a number of changes to the conduct of operations in early 2012 in response to these recommendations, and also promulgated these best prac-

---

[16] Lewis and Holewinski, "Changing of the Guard."

[17] The specific shortcoming, like the tactical directive itself, remains classified.

[18] *Civilian Casualty Update Study*, 2012.

tices to tactical forces in Afghanistan to aid their implementation. In addition, the recommendations were shared with training centers back in the United States to be included in predeployment training. Members of the study team then worked with the Center for Army Lessons Learned (CALL) and compiled the main body of a handbook for soldiers, addressing how to reduce civilian casualties during operations.[19] This handbook was shared with forces in theater, as well as with those preparing for future deployments, and it contained tailored guidance and tactics based on specific lessons from actual civilian casualty incidents.

Note that these efforts were done periodically and sequentially, reexamining Afghanistan operations over time to observe how existing measures reduced civilian casualties, and whether there were new issues that also needed to be addressed. These efforts also benefitted from team continuity and expertise, and subsequent teams were able to observe the benefits of tailored guidance. For example, the revised tactical directive issued by General Petraeus in 2010 had a marked positive impact on civilian casualties from airstrikes, and guidance and tactics introduced as a response to root cause analysis often were effective in reducing those contributions to civilian tolls.[20] Also, U.S. SOF observed increased mission success and decreased rates of civilian casualties as a result of identified root causes.[21] At the same time that enemy tactics and the environment changed, it was also necessary to conduct follow-on studies to revisit guidance and tactics in light of subsequent incidents of interest and provide fine-tuning to address new factors as they emerged.

---

[19] *Afghanistan Civilian Casualty Prevention: Observations, Insights, and Lessons*, Handbook No. 12-16 (Fort Leavenworth, KS: U.S. Army, CALL, 2012), https://publicintelligence.net/call-afghan-civcas/. Members of the study team—LtCol Randolf C. White (USA) and Larry Lewis—coauthored the chapter on consequence management, "What to Do When Civilian Casualties Occur," with Sarah Holewinski and Marla Keenan from the Center for Civilians in Conflict. The center is a nongovernmental organization that advised ISAF and U.S. forces on improving civilian harm mitigation during operations.

[20] Lewis, *Civilian Casualties: Enduring Lessons*.

[21] Lewis, *Reducing and Mitigating Civilian Casualties*.

CHAPTER 16

IMPLEMENTING THE APPROACH

The process outlined and illustrated above quantifies how well operations meet stated U.S. goals of lethal action operations, as well as identifies specific areas of improvement. This review also identifies root causes, providing not only an explanation for why mission success was not achieved or civilian harm occurred, but also a basis for evidence-driven solutions to improve the performance of future operations.

In addition to following the methodology mentioned earlier, the process described here would also benefit from several other features. Particularly, the review process described here should be an independent one—that is, not conducted by a specific agency or Service. Also, the review process should provide feedback to the forces conducting operations to help them improve their own after action processes and promote learning. The review process should pursue ways to improve assessments of civilian casualties during operations as well, given past difficulty to accurately estimate civilian casualties in these kinds of operations, and there are several factors that can lead to official estimates being too low.[1]

---

[1] See Lewis, *Drone Strikes in Pakistan*, for a detailed discussion of this point.

REVIEW SHOULD BE INDEPENDENT

Many sources discuss the existence of at least two different groups for U.S. lethal force operations, one within the military and the other within an OGA. Given the important policy considerations regarding the use of lethal force, each organization is likely to be conducting an internal review of its operations. It would be constructive for these reviews to incorporate elements described here.

However, an independent review process would also be beneficial, especially given the critical nature of these operations and the significant fallout and reduction of freedom of action that can result from negative second-order effects. An independent review, conducted by either an existing organization or a special team of experts assembled for this purpose, would have a number of benefits. First, all organizations have blind spots and can hold assumptions that may not be true. An independent review allows critical analysis of data and assumptions and should yield additional insight regarding these operations. This review can also consider the casualty classification policy, including decisions for individual operations identifying casualties as either combatants or civilians. For example, a similar process conducted for CT operations in Afghanistan yielded both additional understanding of actual root causes of civilian casualties and refinements to casualty classification processes.[2]

Another reason for an independent review is the fact that there are multiple action arms in the U.S. government involved in these operations, each with differing oversight processes. The National Security Council, Congress, the Intelligence Community, and the military all play key roles in shaping, executing, and validating U.S. counterterrorism policy.[3] Yet details regarding operations and lessons often become lost in each group's information silos. An independent review will more effectively identify key lessons

---

[2] *Civilian Casualties in Afghanistan*, TF 3-10 (Suffolk, VA: JCOA, 2012).

[3] The U.S. Intelligence Community (IC) represents a federation of 17 different government agencies that work individually and collectively on intelligence activities to support national security. For more information on the individual agencies and the IC, see http://www.intelligence.gov.

across these organizations, promote consistency, and foster cross-pollination of best practices and lessons. Again, in Afghanistan, such sharing of best practices and lessons among different U.S. units and elements did not occur regularly, and independent reviews that examined operations by different organizations were the most effective way to leverage learning and avoid stovepipes.[4]

REFINE AND STANDARDIZE AARS

As part of its work, the independent review should also examine the sufficiency of AARs produced following each operation. The review should examine such questions as:

- Are the reports factually accurate?
- Are they well suited for operational learning?
- Do they adequately support consequence-management activities?
- What are potential areas of improvement to better support the overall process?

In Afghanistan, U.S. forces and ISAF created a number of products that reported on civilian casualties; the most valuable reports were generally command-directed investigations. These investigations generally considered all available data and interviewed all U.S. government personnel involved in the incident. This process provided a rich dataset for analysis of causal factors, including those that had not surfaced previously or had not been seen as significant when considering a single incident but, when analyzed collectively, emerged as a common contributing factor for many incidents.

Although they were usually the best data source available, these command-directed investigations were not optimal. The purpose of the investigations was to determine culpability, and the focus on learning from the incident was often lost. At the same time, the disparity between the kinds of

---

[4] *Adaptive Learning for Afghanistan: Final Recommendations* (Suffolk, VA: JCOA, 2011), https://info.publicintelligence.net/JCOA-ALA-Afghanistan.pdf.

issues and the data considered became apparent, since investigations were often conducted by different units and individuals. This limited the consideration of some root causes in the analysis. Identifying key issues and information to be included in AARs would address these concerns; for example, a checklist based on the Joint Civilian Casualty Study and other related studies is included in the Army's *Civilian Casualty Mitigation* report.[5] The independent review should develop a tailored set of reporting requirements to ensure AARs best support the learning process for the specific type of operations.

INCREASE THE ROBUSTNESS OF CIVILIAN CASUALTY ASSESSMENTS

In addition to reporting requirements, standard processes should be put in place to improve the accuracy of civilian casualty assessments. The potential for civilian harm does not mean that the engagement is not permissible. Under U.S. and international humanitarian law (e.g., the Geneva Conventions), the use force is permissible against an enemy as long as the harm to civilians is not excessive relative to the gained advantage from the operation. However, civilian tolls should be properly acknowledged in follow-on reporting and assessments.

The United States faces considerable challenges in obtaining accurate assessments of civilian casualties in its use of lethal force in operations outside of declared areas of hostilities. Most of these operations include airstrikes in areas without U.S. boots on the ground and, as such, are often characterized by air-based target identification and battle damage assessment. These factors increase the likelihood that civilian casualties, including those misidentified as enemy combatants, are not discovered by the U.S. government. Thus, if civilian casualty assessment depends only on these measures, the government will likely never have a true picture of the actual scale of civilian harm from its drone strikes. Regarding operations in Pakistan and Yemen, the nation has frequently denied the extent of civilian casualties widely reported in the media. This situation resembles Afghanistan prior to mid-2009, where U.S.

---

[5] *Civilian Casualty Mitigation*, ATTP 3-37.31 (Washington, DC: Department of the Army, 2012), appendix B.

and international force military commanders were frequently confronted by reports of civilian casualties that differed from their own initial reports.[6]

This tendency to underestimate the levels of civilian casualties also undermines the forces' ability to reduce civilian casualties. If the magnitude of civilian harm is underestimated, then the risks are not being prioritized appropriately and causes will not be understood; as a result, measures put in place to reduce civilian harm may not be effective. Therefore, the government can only truly minimize civilian harm with an accurate assessment process to quantify levels of civilian harm and with an analytic process in place to capture root causes showing why civilians are harmed in operations.

Part III provides such an analytic process, but it should be accompanied by an improved assessment process to quantify civilian harm. In addition to intelligence efforts (e.g., use of HUMINT and imagery) to improve assessments, this process could consider information from third-party organizations, either with a presence on the ground or with processes for gathering information that are complementary to that of official U.S. methods. While information gained here may appear to conflict with operational data, such a process can highlight cases where findings from operational data are not as incontrovertible as they seemed, leading to revised estimates of civilian tolls.

One overarching principle of IHL is that "In case of doubt whether a person is a civilian, that person shall be considered to be a civilian."[7] America is governed by this principle with respect to the use of force, but the same principle applies for assessments of civilian harm after the use of force. In cases where a doubt exists as to whether a casualty was a combatant or noncombatant, they should be assigned to noncombatant status per IHL.[8] This was the practice of U.S. military forces in Afghanistan; doing otherwise is inconsistent with international law and is likely to lead to an incomplete picture of the civilian toll from these strikes.

---

[6] Barbara Starr, "U.S., Afghanistan Differ on Number of Civilian Casualties in Strikes," CNN News, 9 May 2009.

[7] *Protocol Additional to the Geneva Conventions of 12 August 1949.*

[8] Author communication with ICRC, May 2014.

CHAPTER 17

BENEFITS OF THE PROCESS

The analytic approach proposed in Part III seeks to improve the conduct of the U.S. CT campaign of lethal force by providing improved context for operations, revealing causes for negative outcomes, and creating tailored recommendations for improved conduct of future operations. This approach also yields several other benefits, including a basis for "defragmenting" the execution and oversight of lethal action across U.S. departments and agencies, separating fact from opinion regarding operations, making the case for broader changes to the U.S. military in light of identified lessons, and enhancing the perceived legitimacy of the U.S. lethal action campaign by better aligning policy and practice.

IMPETUS FOR REVISING GUIDANCE AND OPERATIONAL APPROACHES

U.S. counterterrorism lethal action operations are aimed at imminent threats where no other solution for addressing them—such as working with partner nations or attempting capture operations—is feasible. For this critical mission set, a missed opportunity—or killing the wrong people—can have dire consequences. The analytic process described here may accelerate and optimize the normal learning process to maximize success and minimize the potential for these negative outcomes.

Specifically, the benefits of this analytical approach should include:

- refinements to policy and guidance (including a modified PPG if appropriate);
- alternative tactics and operational approaches to improve mission success and reduce civilian casualties, given observed root causes;
- best practices cross-pollinated across organizations;
- shortfalls highlighted in current capabilities;
- adaptations to adversary approaches that complicate lethal action and targeting criteria; and
- a basis for informed policy decisions regarding organizational responsibilities in the future use of lethal force.

In addition, there is continuing discussion in Congress and policy circles of ending the Authorization for the Use of Military Force (AUMF), which is the basis of approvals to conduct operations against al-Qaeda and associated groups.[1] An end to AUMF would require a new legal basis for the use of lethal force. Several commentators have suggested that this new legal basis could include adapting the PPG framework for the targeting process.[2] Getting the guidance right has potential implications for a broader set of lethal action operations beyond the current campaign against al-Qaeda and affiliated groups. And, as stated earlier, creating the optimal guidance for one period of time does not guarantee it remains optimal for the long

---

[1] ". . . the President is authorized to use all necessary and appropriate force against those nations, organizations, or persons he determines planned, authorized, committed, or aided the terrorist attacks that occurred on September 11, 2001, or harbored such organizations or persons, in order to prevent any future acts of international terrorism against the United States by such nations, organizations or persons." Authorization for Use of Military Force, Public Law No. 107-40 (2001).

[2] See, for example, Harold Hongju Koh, "Authorization for Use of Military Force After Iraq and Afghanistan" (statement before the Senate Foreign Relations Committee, 21 May 2014), http://www.foreign.senate.gov/imo/media/doc/Koh_Testimony.pdf; and Jack Goldsmith, "Agreeing with Harold Koh on the Need for and Contours of a New AUMF," *Lawfare* (blog), 23 May 2014, http://www.lawfareblog.com/2014/05/agreeing-with-harold-koh-on-the-need-for-and-contours-of-a-new-aumf.

term, so putting a process in place to look for regular opportunities to learn lessons and incorporate them into guidance would improve the U.S. government's ability to deal effectively with future threats.

SEPARATING FACT FROM OPINION

When working to optimize guidance and operational approaches, the effort to rely on facts and data instead of opinion and commonly held assumptions will be critical. These opinions can be incorrect, in effect steering efforts away from addressing root causes and undermining possible improvement.

Early efforts to mitigate civilian casualties in Afghanistan seemed based more on opinions and observations from past operations with no contextualization or systematic examination of the actual incidents in question. As a result, early mitigation tended to be less effective, and resulted in new equipment that created few benefits for addressing major issues. In some respects, this is understandable; in this instance, operating forces were under considerable pressure and lacked both the time and expertise to analyze these issues. These early failures highlight a key reason why General McChrystal welcomed an independent team to assist his efforts, as they could come into theater to collect data and then have time to think and analyze outside the situation.

Civilian casualties from drone strikes represent another example of the challenge of sorting between fact and opinion. For lethal action operations, the nature of drone strikes has been widely debated. For example, official U.S. statements and other public statements have stressed the "surgical" nature of drone strikes.[3] This opinion is shared by others; for example, it was stressed in a recent Stimson Center report: "Lethal UAV [unmanned aerial vehicle, or drone] strikes frequently have been criticized for their

---

[3] See, for example, Conor Friedersdorf, "Calling U.S. Drone Strikes 'Surgical' Is Orwellian Propaganda," *Atlantic*, 27 September 2012, http://www.theatlantic.com/politics/archive/2012/09/calling-us-drone-strikes-surgical-is-orwellian-propaganda/262920/.

alleged tendency to cause excessive civilian casualties. This criticism has little basis in fact."[4]

Other organizations claim that drones cause more civilian casualties than the U.S. government and others (e.g., academia, think tanks, advocacy groups, and elements of the UN) have claimed, including human rights groups such as the Center for Civilians in Conflict, Amnesty International, and Human Rights Watch, as well as some in academia. In addition, a UN special rapporteur made similar claims.[5] Analysis of open-source data here tends to agree with these claims; for example, drone strike operations in Yemen were almost four times more likely to cause civilian casualties than other types of operations. These different opinions regarding civilian casualties from drone strikes have real impact; they create uncertainty regarding the level of effort the U.S. government should expend in pursuing improvements in this area.

Analysis of operational data can provide valuable insight when opinions differ. For example, contrary to the assertion of the Stimson report, analysis of operational data from Afghanistan showed that drone engagements were 10 times more likely to cause civilian casualties than those by manned aircraft.[6] Thus "their [drones] alleged tendency to cause excessive civilian casualties" is in fact documented in analysis of official U.S. military data from operations in Afghanistan. Furthermore, the Stimson assertion that "this criticism has little basis in fact" is unfounded.[7] This situation—that the Stimson task force made important assertions that were uninformed by and, in fact, contradictory to findings from official U.S. operational data—reinforces the importance of testing assumptions through an analytic process leveraging operational data.

This example also illustrates how incorrect opinions can be formed. The Stimson report justified their position that drones cause minimal civilian casualties by pointing to the precision of drone platforms and mu-

---

[4] Abizaid and Brooks, *Recommendations and Report of the Task Force on U.S. Drone Policy*, 24.

[5] Emmerson, *Promotion and Protection of Human Rights*.

[6] See *Drone Strikes*.

[7] Abizaid and Brooks, *Recommendations and Report of the Task Force on U.S. Drone Policy*.

nitions, and the advantages of the platform in providing persistence and high-quality intelligence.[8] While the descriptions of platform advantages are true, they are also incomplete; the discussion in the Stimson Center report neglects the full set of factors that contribute to civilian casualties during operations. When drone strikes are considered in an operational context, including the distributed nature of drone operations, training contributions, and employed tactics, these factors collectively contribute to why drone strikes can have a greater propensity to cause civilian casualties than other types of engagements, such as strikes from manned aircraft.[9]

The analytic process described here may identify cases where common opinion differs from fact. This is also important when developing tailored solutions that address the actual drivers for challenges. For example, from analysis of root causes of civilian casualties during drone strikes in Afghanistan, the factors that contributed most significantly could be mitigated—so, this propensity to cause civilian casualties is not an inherent limitation of drones, but rather a situation that can be improved through deliberate measures informed by root causes.

"DEFRAGMENTATION" OF U.S. COUNTERTERRORISM EFFORTS

With regard to the use of lethal force to deal with threats to the United States, many elements of the government have roles and responsibilities in executing and overseeing this program. For example, the National Security Council, Congress, the Intelligence Community, and the military all play key roles in shaping, executing, and validating U.S. CT policy. At the same time, details regarding operations and lessons are often trapped in organizational silos. This issue stems largely from the fact that two groups currently execute lethal action, and they execute on the basis of differing legislative authorities (Title 10 versus Title 50), answering to various congressional committees.

---

[8] Ibid.

[9] This point is discussed in detail in Lewis, *Drone Strikes in Pakistan*.

Having OGAs executing operations introduces barriers to sharing best practices and reduces collective learning. For example, culture, classification and access, doctrine, and equipment create differences in the conduct of operations and make it more difficult to share feedback and enable learning across organizations. This challenge also was seen in Afghanistan; individual units learned in isolation, and key lessons were often not shared from one unit or force to another. Likewise, having contradictory oversight organizations for the two action arms limits their ability to compare the conduct of operations and identify opportunities for improvement.

Given these barriers, an independent review will create an opportunity to more effectively identify key lessons across these stovepipes. By looking at operations across different organizations, key lessons can be identified more effectively and offer the chance to share best practices used by one agency but not the other. These lessons should also be shared across the organizations responsible for oversight to best understand current effectiveness in lethal operations, and what steps need to be taken to address key challenges.

IMPETUS FOR BROADER INSTITUTIONAL CHANGE

President Obama indicated that the use of lethal force in CT operations outside of declared areas of hostilities would shift to the military in the near future.[10] Notably, current U.S. lethal action operations are similar to operations in Afghanistan in that existing guidance for protecting civilians exceeds the baseline legal requirement. Per U.S. and international law, harm to civilians is permissible during military operations if the engagement focuses on a valid military target, if it discriminates between combatants and civilians, and if the use of force is proportional to the threat. The standard for protecting civilians in the PPG—that operations will be conducted only if there is a relative certainty that civilians will not be harmed—goes above and beyond that required to comply with IHL; this

---

[10] Ken Dilanian, "CIA Winds Down Drone Strike Program in Pakistan," *Military.com* (blog), 30 May 2014, http://www.military.com/daily-news/2014/05/30/cia-winds-down-drone-strike-program-in-pakistan.html.

higher standard can be referred to as "supercompliance." If this higher standard of supercompliance continues to be an expectation for the use of force—and recent operations in the past decade suggest this will be the case—then the U.S. military should consider institutional changes to reflect this standard.

Fortunately, the United States has already made some changes to reduce civilian casualties over the past decade. For example, while the government consistently met the IHL requirements in combat operations during this period, it also found that more could be done to avoid civilian casualties. U.S. forces in Iraq and Afghanistan found that their operations could be improved by protecting civilians—and was in fact required for the success of the campaign—over the course of those lengthy operations. However, in these cases, improvement was slow; delays in measures taken resulted in unnecessary harm to civilians, as well as harm to the overall campaigns through alienation of the local population, tarnishing of the U.S. reputation, and limited freedom of action.[11]

While progress in reducing civilian casualties and pursuing supercompliance in Iraq, Afghanistan, Libya, and elsewhere is good news, to date, the changes put into place remain largely focused on supporting operations there. Sharing lessons between operations—and institutionalization of those lessons—is less apparent. For example:

- Lessons from Iraq regarding escalation of force (e.g., checkpoint operations) did not appear to migrate to Afghanistan.

- Lessons regarding air-to-ground operations in Afghanistan did not reach NATO participants in Operation Unified Protector in Libya.[12]

---

[11] Lewis, *Reducing and Mitigating Civilian Casualties*.

[12] For more on NATO's role in this operation, see NATO, "Fact Sheet: Operation Unified Protector: Final Mission Stats," 2 November 2011, http://www.nato.int/nato_static_fl2014/assets/pdf/pdf_2011_11/20111108_111107-factsheet_up_factsfigures_en.pdf.

- Lessons for escalation of force did not inform an incident in July 2012, in which a U.S. Navy ship engaged a small fishing boat in the Persian Gulf.[13]

These examples point to a need for a stronger institutionalized approach within the U.S. military to reduce civilian casualties when possible and conform to greater expectations for civilian protection. This approach would benefit from clear military leadership and policy in this key area. For example, a policy-level position in the Office of the Secretary of Defense (OSD) on civilian harm mitigation might focus policy development, encourage consideration of civilian harm in planning, and advocate institutional development of doctrine, tactics, and materiel solutions to reduce civilian harm in operations. In addition, the deliberate analysis of civilian harm in operations could improve the ability of forces to understand current levels of civilian harm in operations and identify ways to minimize it, guiding the actions of the OSD advocate. These changes would benefit CT operations outside of declared areas of hostilities, both under the current AUMF authority and under a new legal basis for action. At the same time, these measures would yield benefits in a range of other kinds of operations, including the recent airstrikes against ISIL in Iraq and Syria.

To the extent that OGAs will also be conducting operations using lethal force outside of declared areas of hostilities, those agencies would also benefit from similar measures to institutionalize best practices and policies to reduce civilian harm above and beyond that required by IHL.

ENHANCING U.S. LEGITIMACY AND PROMOTING THE RESPONSIBLE USE OF FORCE

Finally, this process can help build legitimacy for the U.S. lethal action campaign and can serve as a model for other nations that will inevitably

---

[13] Laura Rozen, "U.S. Navy Fires on 'Rapidly Approaching' Boat near Dubai," *Back Channel* (blog), Al-Monitor, 16 July 2012, http://backchannel.al-monitor.com/index.php/2012/07/1185/us-navy-fires-on-motor-boat-off-dubai-uae/.

use lethal actions to address their own national interests. The United States regularly advocates for the discriminate use of force and the protection of civilians in conflict in the international community. However, the CT lethal action campaign has resulted in criticism of U.S. policies and practices from an unlikely collection of sources, including elements of the UN, members of the British Parliament, China, academia, NGOs, and other human rights organizations. This broad criticism has arguably injured the moral authority of the nation and degraded its ability to exert global leadership.[14]

A U.S. commitment to this review sends a message that its actions are consonant with its words—that it is acting decisively to protect its citizens from imminent threats while also doing "everything possible" to protect civilians from harm. The Obama administration and Congress can use this review to improve transparency; the government could share broad trends from this review with the public, as well as communicate the benefits gained through the review—namely, specific improvements to promote mission success and progress in reducing civilian harm from CT operations. By conducting this review, the United States can point to this framework and process, giving the country greater credibility when advocating for other nations to follow similar standards in the responsible use of force.

---

[14] For example, Gen James E. Cartwright (USMC), former vice chairman of the Joint Chiefs of Staff, testified to a Senate committee concerning drones: "I am worried that we have lost the moral high ground." See David Lerman, "Democrat Seeks Transparency on Drones as White House Skip," Bloomberg, 24 April 2013, http://www.bloomberg.com/news/articles/2013-04-23/democrat-seeks-transparency-on-drones-as-white-house-skip; and Tom Curry, "Poll Finds Overwhelming Support for Drone Strikes," *NBC Politics* (blog), NBC News, 5 June 2013, http://nbcpolitics.nbcnews.com/_news/2013/06/05/18780381-poll-finds-overwhelming-support-for-drone-strikes?lite.

CHAPTER 18

CONCLUSIONS AND RECOMMENDATIONS

The use of lethal force remains a key component of the U.S. CT approach given the ongoing security threats from al-Qaeda and associated forces. Recent guidance, including the PPG, seeks to balance mission success and the risk of civilian casualties. Having a robust analytical framework and lessons learned process, as outlined here, could quantify how well operations are meeting stated U.S. goals for lethal action operations, including a report card to summarize overall progress, as well as to identify specific areas where improvement is possible. This review would also identify root causes, providing an explanation for why mission success was not achieved or civilian harm occurred and providing a basis for evidence-driven solutions to improve performance of future operations. This process may improve mission success—a critical element of U.S. national security—while also reducing civilian casualties, consistent with U.S. principles and policies that value the protection of innocents. Thus, this approach ensures that U.S. practice and policy are aligned.

Such a process could also serve as an independent review. The review process should provide feedback to the action arms conducting operations to help them improve their own AAR process and promote learning. The review process should also pursue ways to improve assessments of civilian casualties during operations since official estimates tend to be too low.

As noted earlier, the United States has a history of adapting and learning lessons in one operation but not applying those hard-fought lessons to other, similar operations.[1] For example, lessons from Afghanistan for civilian protection were not transferred to U.S. and Coalition forces operating in Operation Unified Protector in Libya, nor did they inform U.S. Navy operations in the Persian Gulf in the case of a shooting of an Indian fishing boat. While the process outlined here would improve the conduct of CT operations in Pakistan and Yemen, such a process would also benefit other operations, such as current U.S. military airstrikes in Iraq. For the longer term, creating an institutional focus area for civilian harm within the U.S. military, with a focus on avoiding civilian casualties while promoting mission success, would help preserve these lessons and keep forces from having to relearn them in future operations. Finally, this review could promote U.S. legitimacy and aid the nation's efforts to advocate responsible use of force within the international community, a particularly important consideration with the proliferation of such technologies as armed drones.

RECOMMENDATIONS

The U.S. government should sponsor an independent analysis of U.S. lethal action operations in CT operations, using the analytic approach outlined in the previous chapters. This review should include:

- A team of independent experts who have full access to operational data to review mission success and potential civilian harm during U.S. counterterrorism operations in areas outside of declared theaters of conflict. This review should leverage the analytic process illustrated here, and also reference and draw upon insights from similar reviews conducted for operations in Afghanistan.[2]

---

[1] *Decade of War: Enduring Lessons from the Past Decade of Operations*, vol. 1 (Suffolk, VA: JCOA, 2012).

[2] This review could be a subset of a larger oversight effort as recommended in Abizaid and Brooks, *Recommendations and Report of the Task Force on U.S. Drone Policy*.

- Concrete recommendations for changes to guidance and operational approaches based on identified root causes. A key element to reducing civilian casualties in Afghanistan was analyzing individual incidents and determining causal factors. When these causal factors were considered collectively, they focused efforts for reducing civilian harm to areas that were most productive. This process could easily be replicated for operations outside of declared theaters of conflict, including a review process to determine the causal factors for the incident as outlined earlier. Periodic reviews would identify causal factors across multiple incidents and identify ways to systematically address them in guidance and operational approaches, including future versions of the PPG.

- "Defragmentation" of oversight and lessons-learned processes for U.S. counterterrorism operations. Different elements of the U.S. government have various roles in shaping, executing, and validating its CT policy. For various reasons, however, details concerning these operations tend to be siloed within each organization, limiting operational learning and effective oversight. The independent review should effectively identify key lessons within these stovepipes and share them across the different organizations responsible for execution and oversight of these operations. The government can improve transparency and set an example for the international community by highlighting this effort and sharing broad trends from this review with the public.

The DOD should improve its institutional capability to reduce civilian harm while maintaining mission effectiveness. The military has a history of openly debating the ethical use of force, and it considers compliance with IHL (codified in the Law of Armed Conflict) an integral part of the U.S. profession of arms.[3] That said, DOD could be better organized and resourced to systematically reduce civilian harm in its operations, especially

---

[3] See Rod Powers, "Law of Armed Conflict (LOAC)," *Rules of War* (blog), http://usmilitary.about.com/cs/wars/a/loac.htm.

given the recent trend requiring supercompliance with regard to civilian protection. Key recommendations for DOD include:

- Create a policy-level position in the OSD that focuses on civilian harm mitigation in the conduct of military operations. Civilian harm mitigation is part of the ethical and professional obligation of being a member of the U.S. military, foundational to the profession of arms, yet there is a gap in military leadership and policy in this key area. A policy-level position in OSD should be created to focus on civilian harm mitigation and better enable supercompliance with regard to civilian protection where possible. Such a position could have a role both in current operations and in the institutionalization of lessons for future operations. The ICRC, such international organizations as the UN, and NGOs can interface with this office as the DOD institutional point of contact in addition to its coordination with operational forces.

- Conduct analysis and develop expertise. The deliberate analysis of operations, including the topic of civilian harm in operations, is a relatively new field, with no systematic program for such work and few established experts. However, this aspect of operations is becoming foundational to the ability to use force in a wide range of operations. DOD should develop expertise on the reduction and mitigation of civilian harm and pursuit of best practices with respect to IHL. This resourcing should include support to operational staffs, which typically lack this expertise and analytical capability.

Other agencies employing lethal force should also improve their institutional capability to reduce civilian harm. To the extent that other government agencies will also be conducting operations using lethal force, they would also benefit from measures to institutionalize best practices and policies to reduce civilian harm and pursue supercompliance, similar to those recommended here for DOD.

PART IV

SECURITY AND LEGITIMACY

LEARNING FROM THE PAST DECADE OF OPERATIONS
BY LARRY LEWIS

CHAPTER 19

INTRODUCTION

Conflict and insecurity spread like spilled ink.[1] The United States has returned to Iraq for the third time in 25 years, most recently to confront the emerging threat of ISIL, also present in Syria, Libya, and elsewhere. The United States continues CT operations in Pakistan, Yemen, Afghanistan, and Somalia against al-Qaeda and associated groups, while also supporting Saudi Arabia in its offensive operations in Yemen.[2] Bold attacks from Boko Haram against civilians in Nigeria continue despite U.S. security assistance.[3] In Mali, a government considered a shining example of democracy in Africa fell in a coup after its military was ineffective against yet another Tuareg uprising.[4] Meanwhile, Russia continues to destabilize

---

[1] This sentiment was echoed recently by Robert Kagan: "We're sort of seeing the world order cracking around the edges." See Michael Crowley, "World War O," *Politico*, 23 April 2015, http://www.politico.com/story/2015/04/world-war-obama-117303.html#ixzz3YEDusiYD.

[2] Steve Almasy and Jason Hanna, "Saudi Arabia Launches Airstrikes in Yemen," CNN News, 26 March 2015, http://www.cnn.com/2015/03/25/middleeast/yemen-unrest/.

[3] Adam Nossiter, "Boko Haram's Civilian Attacks in Nigeria Intensify," *New York Times*, 6 July 2015, http://www.nytimes.com/2015/07/07/world/africa/boko-haram-intensifies-attacks-on-civilians-in-nigeria.html?_r=0

[4] Andy Morgan, "What Do the Tuareg Want?," Al Jazeera, 9 January 2014, http://www.aljazeera.com/indepth/opinion/2014/01/what-do-tuareg-want-20141913923498438.html.

Ukraine, a U.S. partner.[5] Overall, the world is a turbulent place and the U.S. government faces a prodigious number of threats to its security and interests.

The remaining four chapters of this work describe how security challenges over the past decade are often a symptom of a bigger issue: inadequate consideration of *legitimacy* within the overall U.S. calculus of national security. Legitimacy involves the right of a government to govern and exercise such functions as the use of force, which can span from the use of armed drones to tactical checkpoint operations. This right stems from both external and internal considerations, and can be broken into two parts: international and domestic legitimacy. International legitimacy has been defined as ". . . a measure of the acceptability and justifiability of a state's actions in the eyes of other states and their citizens."[6]

A key consideration for a government's international legitimacy is adherence to international norms, including, as discussed here, norms for armed conflict. Likewise, an indicator of the domestic legitimacy of a government is the degree to which a nation's population accepts its political authority, and ". . . is usually related to the achievement of social and distributive justice and thus revolves around the existence of a government *for* the people."[7] Total legitimacy is not always required to rule or to use force in an armed conflict; however, legitimacy promotes the ability to govern or influence without resorting to authoritarian regimes or repressive force.[8] Lack of legitimacy undermines efforts to govern and maintains a monopoly on the use of force. Likewise, factors that reduce legitimacy (e.g., heavy-

---

[5] "Stoltenberg: Russia Continues to Destabilize Situation in Ukraine," Ukrinform, 17 April 2015, http://www.ukrinform.ua/eng/news/stoltenberg_russia_continues_to_destabilize_situation_in_ukraine_330713.

[6] Suzanne Nossel, "Going Legit," *Democracy: A Journal of Ideas* 3 (Winter 2007): 29–38, http://www.democracyjournal.org/magazine/3/going-legit/.

[7] Jean D'Aspremont, "Legitimacy of Governments in the Age of Democracy," *Journal of International Law and Politics* 38 (2007). Emphasis in original.

[8] See ". . . legitimacy is a functional prerequisite of efficient and liberal forms of government." in Fritz W. Scharpf, *Reflections on Multilevel Legitimacy*, Working Paper 07/3 (Cologne, Germany: Max Planck Institute for the Study of Societies, 2007), 7.

handed approaches, corruption, etc.) also tend to serve as grievances for extremist or guerilla groups, undercutting security.

In recognition of its importance, U.S. military doctrine includes legitimacy in its list of "principles of Joint operations."[9] This list constitutes a set of best practices that should be followed for any conflict. While it is commendable to find this in *Joint Operations* doctrine, it is less apparent in policy and practice for the United States. Over the past decade, there has been an apparent lack of consideration for legitimacy in both U.S. government activities related to armed conflict as well as larger policies and practices aimed at dealing with existing threats, which encompass both combat operations and security force assistance. As a result, the United States dealt with security concerns less effectively, employing incomplete solutions and conducting tactical actions that, at times, undermined larger strategic objectives. In the following chapters, we will discuss how this has contributed to conflict and insecurity.

AN OPPORTUNITY TO LEARN

"Failure is only the opportunity more intelligently to begin again."

~ Henry Ford[10]

A few years ago, General Dempsey, as chairman of the Joint Chiefs of Staff, called for the U.S. military to "learn the lessons from the past decade of operations." A study conducted by the Joint Staff J7, with a product commonly referred to as the *Decade of War* report, was a key element of that process.[11] Written in 2012, that report was optimistically called volume one in recognition that its scope was by no means exhaustive; more effort would be needed to address additional issues not touched on in the first volume. While a second volume has yet to be released, there are challenges where the U.S. government urgently needs to learn key lessons.

---

[9] *Joint Operations*, Joint Publication 3-0 (Washington, DC: Joint Chiefs of Staff, 2011).

[10] Henry Ford and Samuel Crowther, *My Life and Work* (New York: Doubleday, 1922), 19.

[11] *Decade of War*.

The 2012 report summarized key insights from a collective body of work regarding lessons from U.S. military operations. For the primary author of that report, it became evident that this same body of work, combined with other, more recent reports, offers lessons for the U.S. government on the topic of legitimacy and national security. These lessons are two-fold: a set that includes considering the legitimacy of other nations in solving national security concerns, and a set for the nation and its own conduct of operations relating to legitimacy. The second set of lessons is critical to the first, as the United States cannot effectively promote legitimacy within other nations if its own foundation is unsound. Key lessons for these two areas will be addressed in turn.

CHAPTER 20

LESSON ONE: PROMOTING LEGITIMACY

As stated above, nations have two dimensions that contribute to legitimacy: adherence to law and the equitable governance of its citizens. With respect to adherence to law in a security context, international humanitarian law (IHL) is particularly relevant. IHL is often referred to in the United States as the law of armed conflict (LOAC).[1] IHL aims to "ensure a degree of humanity in the midst of war" by setting minimum standards for behavior in combat, as well as prescribing steps to minimize the extent of human suffering with regard to the use of force.[2] Legitimacy of a government also includes governance that is fair, relatively effective, and transparent.

The importance of these two factors can be seen in many examples from U.S. operations over the past decade. For example, sectarian tendencies, including selective governance to only parts of the population as well as such practices as the torture of Sunni Iraqis at the hands of Shiite militia and Iraqi security forces, undermined the legitimacy of the post-2004 Iraqi govern-

---

[1] The body of IHL primarily consists of the four Geneva Conventions of 1949 and the two Additional Protocols of 1977.

[2] Carlo von Flüe and Jean-Philippe Lavoyer, "How Can NGOs Help Promote International Humanitarian Law?," *Humanitarian Exchange*, no. 9 (November 1997).

ment.[3] This lack of legitimacy led to unresolved grievances that contributed to ISIL's ease in capturing western Iraq last year. In Mali, the Tuareg's periodic uprisings were similarly triggered by long-standing grievances that include selective governance and heavy-handed government responses dating back to the 1960s.[4] Similarly, the Nigerian government is marked by gross human rights violations and corruption, undermining its legitimacy against Boko Haram.[5]

In working with other nations, particularly to address threats to the United States and its interests, the consideration of human rights and legitimacy can take a back seat to the expediency of reestablishing security. With this emphasis on security, U.S. policy and practice can neglect the often causal tie of human rights and the related issue of host nation legitimacy to the emergence and strength of such threats. By putting less priority on promoting human rights and governance to encourage legitimacy with nations facing security threats early on, more effort and resources may be required later. By then, threats are more mature and capable, the population may be more likely to lend support to nonstate armed groups, and U.S. military responses—in terms of cost, represents the big ticket item for American policy options—can become necessary.

EXAMPLE OF SECURITY ASSISTANCE CHALLENGES

The connection between human rights, legitimacy, and security can be seen in an example where an armed group rose up to present an existential threat to a nation. In this case:

- the government had marginalized a minority population group in terms of participation and services for years;

---

[3] See *After Liberation Came Destruction: Iraqi Militias and the Aftermath of Amerli* (New York: Human Rights Watch, 2015), http://features.hrw.org/features/HRW_2015_reports/Iraq_Amerli/index.html.

[4] Morgan, "What Do the Tuareg Want?"

[5] See, for example, Nossiter, "Boko Haram's Civilian Attacks in Nigeria Intensify"; and Natalie Chwalisz, "Recent Hearing: The Continuing Threat of Boko Haram," *Security Assistance Monitor*, 20 November 2013, https://securityassistancemonitor.wordpress.com/2013/11/20/recent-hearing-the-continuing-threat-of-boko-haram/.

- the military employed indiscriminate use of force and denied justice to that group, undermining the force's legitimacy and leading to grievances that fueled the conflict;

- a neighboring country offered sanctuary, allowing the group to better train and gain resources;

- the group benefitted from seasoned military leadership, despite being marginalized within the nation's security forces;

- the group acquired modern equipment by capturing it from an unsound but U.S.-supported government force;

- the resultant uprising was a repeat occurrence; while a previous conflict was quelled, the underlying factors leading to conflict were not addressed; and

- the collective result of these factors was an armed group that, with only hundreds of personnel, was able to defeat a national security force of thousands, capturing territory and equipment.

These factors can easily describe ISIL's dramatic capture of Mosul in June 2014, which served as a wake-up call to many nations regarding the threat of ISIL.[6] However, the factors equally apply to the Tuareg capture of northern Mali in 2012, which precipitated a government coup. Mali illustrates several shortfalls of the overall U.S. approach to security threats through security assistance. America can expend considerable resources on security assistance for weak states, but these expenditures often focus on two areas: tactical training/education and providing military equipment. While these activities are often appreciated by the receiving state, they do not necessarily translate into improved capability or capacity. Furthermore, when these efforts are made without considering the soundness of the U.S. approach in relation to the state's legitimacy, capacity, and ownership of the focus areas

---

[6] Julia McQuaid et al., *Adaptive and Innovative: An Analysis of ISIL's Tactics in Iraq and Syria* (Alexandria, VA: CNA, 2015).

of security assistance efforts, government support can prove futile, as was the case in Mali.

For more than a decade, Mali was viewed by U.S. and international observers as a "democratic success" in Africa.[7] At the same time, Mali faced several key challenges, including scarce resources and abject poverty, with 44 percent of the population falling below the poverty line.[8] Also, northern Mali hosted a terrorist threat—al-Qaeda in the Islamic Maghreb (AQIM)— which threatened regional stability and U.S. interests.[9] Mali's security challenges were compounded by a long-standing Tuareg insurgency in the north. Accordingly, Mali received sizable amounts of U.S. funding, both for development aid and security assistance to their military to enhance their CT capabilities. However, when a military coup in 2012 overthrew Mali's democratically elected government—the coup was led by a Malian soldier who received U.S. training and attended a U.S. military school—the effectiveness of U.S. assistance and aid in promoting stability in Mali came into question.[10]

There were underlying flaws in the U.S. approach to support in Mali that doomed the effort. These flaws included both host-nation considerations and features of the American government's approach to building

---

[7] Alexis Arieff, *Crisis in Mali* (Washington, DC: Congressional Research Service, 2013), https://www.fas.org/sgp/crs/row/R42664.pdf.

[8] "Mali," World Bank, accessed 18 November 2013, http://data.worldbank.org/country/mali. Also, Mali was rated 182 out of 187 countries in the 2012 UN Human Development Index. See *Human Development Report 2013* (New York: United Nations Development Programme, 2013), http://hdr.undp.org/en/2013-report.

[9] For more information on this Salafi-jihadist militant group, see Zachary Laub and Jonathan Masters, "Al-Qaeda in the Islamic Maghreb," Council on Foreign Relations Backgrounder, 27 March 2015, http://www.cfr.org/terrorist-organizations-and-networks/al-qaeda-islamic-maghreb-aqim/p12717.

[10] Adam Nossiter, "Soldiers Overthrow Mali Government in Setback for Democracy in Africa," *New York Times*, 22 March 2012, http://www.nytimes.com/2012/03/23/world/africa/mali-coup-france-calls-for-elections.html?_r=0.

partner capacity (BPC) in Mali. Host-nation considerations that were not adequately considered included:[11]

- **Mali's national ownership:** Mali did not display ownership of its challenges, lacking the political will to devote its resources and efforts to address key challenges. Mali also did not use U.S. and international support as intended.

- **Mali's willingness and ability to confront the threat:** the Malian government was not willing or able to confront threats in the northern part of the country. In addition to lack of ownership, this inability to confront the threat came as a result of Malian forces being untrained and the government being unable to sustain its forces so that Mali could govern its territory and secure its borders against emerging threats. As a result, Malian forces were not effective against a much smaller, but well-equipped, insurgent force.

- **Long-standing grievances:** while the uprising in the north and subsequent coup were symptomatic of long-standing grievances held by the population dating back to Malian independence or before, the government had not acted to resolve these issues.

- **The Malian government's lack of perceived legitimacy:** though the United States and other international observers saw Mali as a "democratic success," the government's legitimacy among the population suffered due to corruption and human rights concerns, which aggravated other grievances and contributed to the coup.

In addition, there were features of the American BPC effort in Mali that hindered progress, including the following:[12]

---

[11] These bullets were adapted from a classified report by Richard Moody and Larry Lewis, *Learning from the Crisis in Mali: Lessons for Building Partner Capacity* (Suffolk, VA: JCOA, 2014).

[12] These bullets were adapted from Moody and Lewis, *Learning from the Crisis in Mali*.

- **a lack of unity of effort** within the U.S. government, including disparate strategies, policies, and plans, as well as differences in organizational culture, roles, and missions;

- **a mismatch in national interests, goals, and objectives** between the United States and Mali;

- **inadequate consideration of host nation ownership and legitimacy**, including Mali's political will, capacity, compliance with IHL, and accountability; and

- **a tactical train and equip focus** that neglected institution-building, including considerations of legitimacy.

General Carter F. Ham, the U.S. Africa Command (AFRICOM) commander at the time of the coup, acknowledged this final point: "We were focusing our training almost exclusively on tactical or technical matters. We didn't spend probably the requisite time focusing on values, ethics, and a military ethos."[13]

While host-nation limitations are variables the U.S. government cannot control, the country does have considerable influence. Such influence is often fragmented in practice (as it was in Mali), with U.S. leaders emphasizing different interests.

SUSTAINABLE SECURITY: TWO COUNTEREXAMPLES

The past decade offers two examples where the pattern described above has been changed: the Philippines and Colombia. In both cases, the governments struggled with security concerns and a history of heavy-handed responses to threats with observed gross human rights violations on both sides. In addition, neither government appeared to be providing for its people, offering only limited government services and focusing development opportunities for select groups of the population. As a result, both countries struggled

---

[13] "Mali Crisis: U.S. Admits Mistakes in Training Local Troops," BBC News, 25 January 2013, http://www.bbc.com/news/world-africa-21195371.

with legitimacy within their own populations and internationally. However, both these nations adapted their approach to improve their legitimacy and effectively address serious security threats.

In the Philippines, the United States took on an advise-and-assist mission for more than a decade using U.S. Army Special Forces, with the goal of helping the Philippines fight terrorists. Besides helping Philippine military forces become more tactically proficient, this long-term, on-the-ground partnership led to a number of larger adaptations, including a change of senior leader mindset and emphasis, improved targeting processes that mitigated collateral damage, and the establishment of a Civic Action Group that institutionalized the military's role in protecting the population. The combined effect of these steps successfully broke the cycle of violence and resulted in a dramatic improvement in security.[14]

Similarly, in Colombia in the 1990s, the Colombian security forces, the FARC, and government-sponsored paramilitary forces all committed gross abuses of human rights.[15] With Plan Colombia, Colombia instituted a number of reforms and initiatives to improve its security situation, supported by the U.S. government through extensive security assistance activities.[16] Colombia's early focus was on security and counternarcotics. This objective was aided by such initiatives as the Home Guards program, which improved ties to local communities and promoted more accurate operations to minimize collateral damage.[17] The Colombian government made significant progress in human rights compliance and accountability, and also bolstered its legitimacy among the population through such programs as President Álvaro Uribe's *Plan Nacional de Consolidación* (National Consoli-

---

[14] Geoffrey Lambert, Larry Lewis, and Sarah Sewall, "Operation Enduring Freedom–Philippines: Civilian Harm and the Indirect Approach," *PRISM 3*, no. 4 (September 2012): 117–35.

[15] FARC is an acronym for *Fuerzas Armadas Revolucionarias de Colombia* (Revolutionary Armed Forces of Colombia—People's Army), one of the oldest, largest, and wealthiest left-wing rebel groups in the country.

[16] See "U.S. Policy in Colombia," Amnesty International, http://www.amnestyusa.org/our-work/countries/americas/colombia/us-policy-in-colombia.

[17] *Learning from Colombia: Principles for Internal Security and Building Partner Capacity* (Suffolk, VA: JCOA, 2013).

dation Plan, or more widely known as Plan Colombia).[18] These collective efforts to improve legitimacy reduced popular support for the FARC and other armed groups. Combined with more effective and targeted offensive operations, Colombia now finds itself in a position of strength in ongoing peace talks to end decades of conflict.[19]

These two cases share a number of features. In both cases, the host nation government:

- adapted its security forces for more effective offensive operations;
- improved governance for all, not a select few; and
- focused on proportional use of force and human rights compliance.

These features contributed to the reestablishment of security. The first factor—the use of force to degrade the threat—is an important element and one in which the United States excels, both in its own operations and in advising its partners. The second and third elements—governance and human rights compliance—were also critical as they had both near-term and long-term benefits to security. In the short term, these elements reduced popular support for armed groups and led to improved intelligence for operations. They also were essential to longer-term sustainable security by collectively quelling long-standing grievances among the population and aiding reconciliation, strengthening the legitimacy of the host nation both domestically and internationally.

The American experience with Colombia illustrates a model for effective security assistance, whereby the government provided tactical-level training but also supported Colombian efforts to build institutional capacity. The government, guided by a robust U.S. embassy and military advisors on the ground and embedded within headquarters, also assisted the Colombian military with establishing supportable, niche capabilities for improving operations. All three elements of this approach—tactical-level training, institu-

---

[18] For more information, see "Plan Colombia," Embassy of the United States, Bogotá, Colombia, http://bogota.usembassy.gov/plancolombia.html.

[19] Described more fully in *Learning from Colombia*.

tion building, and providing niche capabilities—included efforts designed to improve Colombian government legitimacy. These efforts included training on the discriminate use of force, improving accountability measures within the military justice system, revising the rules of engagement, and reforming their national judicial system. The United States also supported Colombian initiatives designed to address grievances related to governance and the provision of services in rural areas.

The robust military group (MILGROUP) within the U.S. embassy in Bogotá complemented the efforts of the other U.S. interagency partners, highlighting the benefit of a model where the embassy becomes a command post for synchronizing government efforts. The longer-term, persistent presence of U.S. personnel had other benefits: building strong relationships with Colombian counterparts, and thereby increasing U.S. influence. Key elements of this full spectrum approach—tactical-level training, institution building, and providing niche capabilities—combined with an emphasis on promoting legitimacy and resourcing U.S. embassy-level efforts, would benefit other U.S. security assistance efforts.

Improved unity of effort across the U.S. government is a best practice that is difficult to achieve in practice. The reality of this effort could be seen in the Philippines but was more episodic, depending on personalities at the U.S. embassy. A U.S. government speaking with a single voice to the partner nation is also valuable but equally difficult to achieve. For example, at a key moment in Iraq, during the surge in 2007, a key leader engagement plan was developed for senior Iraqi leaders and all engagements were treated as opportunities to further the objectives in that plan. So, instead of U.S. engagements pursuing different and sometimes conflicting agendas with little forward progress, such engagements aimed to move the host nation in productive directions agreed upon by all government parties. A similar concerted effort across involved elements of the U.S. government in other settings would promote more effective security assistance overall.

These efforts need to be resourced and prioritized in light of their potential impact in promoting long-standing stability and improved national security. All too often, U.S. government budgets generously resource a potential military response but give short shrift to initiatives that could avoid or shorten conflict. In a financially austere environment, more effective security assistance could improve long-term stability and thus curb the need for larger-scale operations that tend to be much more expensive. For example, even the U.S. contribution to Plan Colombia, a significant investment of time and resources, is dwarfed by U.S. expenditures in Iraq and Afghanistan.[20] Importantly, Colombia now exports security and training to its neighbors in the region, a further dividend of U.S. efforts. While this indirect approach may not always be feasible—and direct action in response to security concerns must sometimes be taken—in general, the maxim "an ounce of prevention is worth a pound of cure" appears to suggest a promising and fiscally responsible approach to promoting stability and countering threats.[21]

CIVILIAN CASUALTIES BY PARTNER NATIONS

Another factor impacting legitimacy is the issue of civilian casualties during conflict; for example, during the Israel-Palestine conflict, significant hostilities broke out periodically in the region. The most recent instance includes Israel's conflict with Hamas in 2014. In this conflict, Israel faced a challenging threat in the form of rocket fire launched into its sovereign territory from Gaza, endangering its citizens. Israel used force against Hamas to reduce this threat.[22]

Hamas attacks against Israel are indiscriminate and aimed at civilian targets within Israel, and thus are a violation of international law. However,

---

[20] For example, the total U.S. contribution to Plan Colombia represents about 1 percent (or more than $5 billion since 2000) of the overall cost of U.S. operations in Afghanistan. See "U.S. Policy in Colombia"; and *Learning from Colombia*.

[21] Benjamin Franklin, "On Protection of Towns from Fire," *Pennsylvania Gazette*, 4 February 1735, http://www.historycarper.com/1735/02/04/on-protection-of-towns-from-fire/.

[22] "Gaza Militants Fire Barrage of Rockets into Southern Israel," *Haaretz*, 12 March 2014, http://www.haaretz.com/news/diplomacy-defense/1.579440.

many observers also raised concerns about the high numbers of civilian casualties caused by Israel during the 2014 operations, including UN Secretary General Ban Ki-moon, when UN reports showed that about 7 out of every 10 casualties from Israeli operations were civilian noncombatants. While Israel has contested this number, they acknowledge that about half of the total casualties were civilians.[23]

The United States has a vested interest in civilian casualties caused by Israel given that it provides billions of dollars to Israel for security assistance annually, including weapons, and works diplomatically in the region to reduce conflict and promote long-term stability.[24] While a number of countries and international organizations have voiced concerns about a lack of discrimination in Israeli operations, the U.S. government has been sending mixed messages regarding Israel and civilian casualties. The State Department expressed concerns similar to the UN, including spokeswoman Jen Psaki's comment, "We once again stress that Israel must do more to meet its own standards and avoid civilian casualties."[25] However, General Dempsey, speaking from a DOD perspective, stated that "I actually do think that Israel went to extraordinary lengths to limit collateral damage and civilian casualties."[26] Dempsey also described how a U.S. military team went to Israel to observe Israeli forces and their operations, including lessons learned regarding civilian protection.[27]

---

[23] This ratio is much higher than that observed in recent U.S. operations. See, for example, Lewis, *Drone Strikes in Pakistan*.

[24] Bernadette Meehan, "5 Things You Need to Know About the U.S.-Israel Relationship Under President Obama," *White House* (blog), 1 March 2015, https://www.whitehouse.gov/blog/2015/03/01/5-things-you-need-know-about-us-israel-relationship-under-president-obama.

[25] Nolan Feeney, "U.S. Condemns Gaza School Attack as Israel Says 'Battle Is Ongoing'," *Time*, 3 August 2014, http://time.com/3076108/gaza-israel-un-ban-ki-moon-hamas/.

[26] David Alexander, "Israel Tried to Limit Civilian Casualties in Gaza: U.S. Military Chief," Reuters, 6 November 2014, http://www.reuters.com/article/2014/11/06/us-israel-usa-gaza-id USKBN0IQ2LH20141106.

[27] Notably, no record exists of civilian harm mitigation lessons from this effort in DOD's Joint Lessons Learned Information System (JLLIS), so any such lessons are not being promulgated effectively within the system designated for this purpose.

Assessing Israeli efforts as "extraordinary" seems overstated, however, in light of the best practices developed by the U.S. government to mitigate civilian harm above and beyond IHL requirements over the past decade.[28] These best practices came through difficult lessons; for example, in the early years of Operation Enduring Freedom in Afghanistan, leaders described the U.S. approach as doing "everything possible" to avoid civilian casualties, however, its operational approach did not always reflect this aspiration in practice. While the military generally observed IHL principles of proportionality and distinction, steps still could be taken to better protect civilians. But starting in 2009, the United States made a concerted effort to go above and beyond the requirements of IHL, moving from compliance to supercompliance in the form of comprehensive civilian harm mitigation efforts. Importantly, further analysis demonstrated the operational benefits of this move to supercompliance: targeting effectiveness and friendly force protection were not negatively impacted as civilian casualties were reduced. In fact, for some operations, mission effectiveness improved as civilian casualties decreased. This significant shift included a general consideration of strategic effects from tactical actions: forces not only asking whether they *could* use force, but also whether they *should*, and if so, what tactical alternatives were available to achieve the desired effects while mitigating the impact on the civilian population. In addition, the United States developed its Presidential Policy Guidance concerning CT actions, including drone strikes. This represented the United States making a policy decision to protect civilians beyond the requirements of IHL. The United States also developed evidence-driven tactics for a variety of situations and warfare areas.[29]

---

[28] Though this becomes less surprising when considering the composition of the lessons learned team. The team members from the joint lessons learned organization in J7 were not subject matter experts on the issue of civilian casualties.

[29] Many of these practices and lessons are included in *Afghanistan Civilian Casualty Prevention: Observations, Insights, and Lessons*, Handbook No. 12-16. More detailed lessons and tactics are contained in joint lessons learned reports produced for ISAF and U.S. counterterrorism forces, collectively referenced in Lewis, *Reducing and Mitigating Civilian Casualties*. Note that the team deployed to Israel did not contain any personnel associated with these efforts.

While Israel reports considerable efforts to comply with IHL, it does not seem to have gone through a similar process of pursuing supercompliance. The forces' tactic of roof knocking, for example, where small weapons are used to alert residents that their building is about to be destroyed, appears analogous to the green laser dazzler U.S. forces used in checkpoint situations.[30] They are both methods of warning, but they are both equally flawed in that they do not reliably and effectively protect civilians. The U.S. government observed this deficiency with the laser dazzler and changed its tactics and tools to reduce civilian deaths at checkpoints. Israel should similarly reexamine their tactic and explore other alternatives, as well as creatively improve their processes overall so that they better protect civilians while maintaining effectiveness.

High civilian tolls detract from sustainable security; Israel's tactics harm civilians and continue to fuel resentment among its neighbors. Israel's periodic military operations, used to "mow the grass," match Galula's observation that ". . . conventional operations by themselves have at best no more effect than a fly swatter. Some guerrillas are bound to be caught, but new recruits will replace them as fast as they are lost."[31] America could do more to affect this. Currently, U.S. engagements and public remarks concerning Israel are inconsistent, reducing the government's ability to influence Israel to learn relevant lessons from their own experiences while offering related lessons from the U.S. experience in Afghanistan and elsewhere. By unifying U.S. engagements and pushing for learning and improvement in Israel's efforts to reduce civilian casualties and improve its

---

[30] See, for example, Adam Taylor, "'Roof Knocking': The Israeli Military's Tactic of Phoning Palestinians It Is about to Bomb," *Washington Post*, 9 July 2014, https://www.washingtonpost.com/news/worldviews/wp/2014/07/09/roof-knocking-the-israeli-mjilitarys-tactic-of-phoning-palestinians-it-is-about-to-bomb/; and Patrick Tucker, "What Will Happen to You When You Storm a U.S. Military Checkpoint?," Defense One, 15 July 2015, http://www.defenseone.com/technology/2015/07/what-will-happen-you-when-you-storm-us-military-checkpoint/117898/.

[31] "Mowing the grass" refers to periodic combat operations to temporarily reduce threat levels, while doing little to keep the resurgence of the threat down in the future. See, for example, Daniel Byman, "Mowing the Grass and Taking Out the Trash," *Foreign Policy*, 25 August 2014, http://foreignpolicy.com/2014/08/25/mowing-the-grass-and-taking-out-the-trash/; and David Galula, *Counterinsurgency Warfare: Theory and Practice* (Westport, CT: Praeger, 1964), 51.

legitimacy, the U.S. government might reduce grievances and break the pattern of long-standing security challenges in the region, promoting Israel's security against continuing attacks, as well as U.S. strategic interests.

Providing security assistance—and specifically the provision of weapons—to nations that then create high civilian tolls with their operations may make the American government appear complicit in the resulting humanitarian concerns. For example, providing bombs to Israel while the UN and other international observers raise concerns about Israel's lack of discrimination and proportionality in airstrikes affects the perception of U.S. legitimacy. A similar situation arose with Saudi Arabia's 2015 airstrike campaign in Yemen.[32] With the United States providing weapons, operational support, and intelligence to Saudi Arabia, the civilian casualties from their airstrikes reflect back on the government. And, in cases where the United States is providing materiel or operational assistance, these actions carry an implicit responsibility for influencing the use of those resources. This represents not only a legal responsibility under IHL, but also a concern for promoting U.S. legitimacy and interests.[33] The U.S. government can take a number of steps to promote civilian protection as part of ongoing security assistance programs, including specific tools for reducing civilian harm, training and mentoring in civilian harm mitigation, and tracking and analyzing civilian casualties. Overall, these actions should both promote legitimacy and reduce grievances, thus improving long-term security.

---

[32] Mohammed Tawfeeq and Dana Ford, "Saudi Arabia Launching Political Solution Campaign in Yemen," CNN News, 21 April 2015, http://www.cnn.com/2015/04/21/middleeast/yemen-crisis/.

[33] The American government has two reasons for promoting IHL compliance by other nations under international law: (1) under common Article 3, the United States is considered a party to the conflict, representing cases where the government provides materiel support to operations; and (2) under common Article 1, with the general requirement for promoting IHL compliance by other nations. This general requirement is particularly relevant for partners where the U.S. government provides security assistance.

CHAPTER 21

LESSON TWO: PRACTICING LEGITIMACY

A key element of U.S. legitimacy in the context of national security is conformity of U.S. operations to international and domestic laws, particularly IHL, also known as LOAC.[1] IHL governing the conduct of armed conflict falls into two categories: the conduct of hostilities (e.g., the means and methods of fighting, including protecting civilians) and detention (e.g., how captured personnel are treated). Over the past decade, domestic and international observers have voiced concerns over both aspects of U.S. activities. Concerns over the U.S. conduct of hostilities include prohibited treatment of medical facilities (e.g., making them military objectives or not observing their protected status); such practices as signature drone strikes in CT operations (where the U.S. government acknowledges that it does not have a good understanding of who was being targeted); and incidents that caused civilian casualties during U.S. counterterrorism operations in Pakistan and Yemen, as well as counterinsurgency operations in Iraq and Afghanistan. While not the focus of this discussion, U.S. treatment of detainees raises concerns about abuses in Guantánamo Bay Detention Center, Abu Ghraib prison and other facilities in Iraq, Bagram in Afghanistan, the CIA's use of enhanced interrogation techniques—some of which are widely

---

[1] The body of IHL primarily consists of the four Geneva Conventions of 1949 and the two Additional Protocols of 1977.

regarded to constitute torture—holding detainees secretly in various U.S. facilities in Guantánamo and other locations around the world, and extraordinary rendition practices.[2]

CONDUCT OF HOSTILITIES

The United States has long been committed to upholding IHL with regard to the conduct of hostilities, including minimizing collateral damage that encompasses civilian casualties (both deaths and injuries) and unintended damage to civilian objects (e.g., facilities, equipment, or other property not considered a military objective). In support of these goals, the U.S. military developed capabilities for precision engagements and accurate identification of targets, including developing refined targeting processes and predictive tools to better estimate and minimize collateral damage. These capabilities enabled combat operations with fewer civilian casualties. Despite these efforts, and while maintaining compliance with IHL, the U.S. military found over the past decade that these measures were not always sufficient to meet the goal of minimizing civilian casualties. The resulting civilian casualties ran counter to American desires and strategic objectives, seemingly contradicting public statements that the United States did "everything possible" to avoid civilian casualties, and therefore caused negative second-order effects that impacted U.S. national, strategic, and operational interests.[3] The rest of this chapter will discuss civilian protection in initial combat operations in Afghanistan and Iraq, counterinsurgency operations in those countries, and CT operations outside of declared theaters of conflict.

---

[2] Extraordinary rendition is the transfer of a detainee to a foreign country for the purposes of detention and interrogation without following legal due process.

[3] International Security Assistance Force (ISAF), "Operational Update 06 February 2012" (press release, Kabul, Afghanistan, 6 February 2012), http://www.rs.nato.int/images/stories/File/06%20Feb%2012%20ISAF%20spokesperson%20Press%20Statement.pdf.

AFGHANISTAN, 2001–2

The United States reaffirmed its commitment to minimizing harm to civilian populations when it commenced major combat operations in Afghanistan and Iraq in 2001 and 2003, respectively. On 7 October 2001, U.S. forces began combat operations to capture al-Qaeda leadership and eliminate Afghanistan as a launching point for terrorism. Within days of the start of operations, international media reported incidents of civilian casualties. Many of these incidents involved villages where suspected enemy fighters were located, highlighting the challenge posed by fighting an enemy that eschews its obligations under IHL (e.g., not wearing a uniform and hiding among the civilian population). As a result, obtaining positive identification was more problematic, and U.S. engagements relied more on self-defense considerations based on perceived hostile acts or intent. Probably the two highest profile incidents during this time were the 21 December 2001 U.S. air attack on a convoy the Afghan government claimed to comprise tribal leaders; and a 1 July 2002 U.S. airstrike on a group that was revealed to be a wedding party in Deh Rawud, in central Afghanistan. In both attacks, U.S. aircraft had observed ground fire and engaged because of self-defense considerations.

MAJOR COMBAT OPERATIONS IN IRAQ, 2003

In major combat operations in Iraq, the ability to distinguish the enemy from the civilian population was simplified by the fact that the enemy was the Iraqi military. Iraqi forces were generally located apart from civilian areas; their military equipment and uniforms reduced the ambiguity of engagement decisions relative to those faced by U.S. forces in Afghanistan. However, the Iraqi military purposely violated IHL rules designed to protect the peaceful civilian population by employing human shields, misusing protected symbols for humanitarian organizations (e.g., the Red Crescent), and placing equipment in protected sites. In addition, *Saddam Fedayeen* (or Men of Sacrifice) forces did not wear uniforms and fought using irregular tactics, contributing to challenges obtaining positive identification for strikes. In contrast, the United States and its allies went to great lengths to minimize collateral damage; for example, in Iraq, similar to Afghanistan, most air engagements used precision-guided munitions. While DOD assessments of civilian casualties during major combat operations in Iraq do not appear to have been performed, an independent assessment judged U.S. preplanned attacks to be relatively effective in minimizing civilian casualties. The main concerns over civilian casualties centered on Coalition forces conducting time-sensitive targeting of leadership in urban areas.[4]

COUNTERINSURGENCIES IN IRAQ AND AFGHANISTAN[5]

As insurgencies developed in Iraq and Afghanistan, the United States was forced to adopt a counterinsurgency (COIN) approach in these countries for which it was largely unprepared. With civilian protection being a central feature of COIN, the reduction and mitigation of CIVCAS (civilian casualties) became a key issue in these operations.

COIN in Iraq

In Iraq, civilian casualties were primarily caused by escalation of force (EOF) incidents, both at checkpoints and during convoy operations. This resulted in significant outcry from nongovernmental organizations (NGOs) and the media; the attack on a vehicle containing Italian journalist Giuliana Sgrena and her rescuers during an EOF incident further increased visibility of this issue.[6] In mid-2005, U.S. forces in Iraq adapted and heightened efforts, which were widely seen as successful, to prevent and mitigate the effects of CIVCAS.[7] Still, this issue was not completely resolved; later in

---

[4] *Off Target: The Conduct of the War and Civilian Casualties in Iraq* (New York: Human Rights Watch, December 2003).

[5] This section was adapted from Lewis, *Reducing and Mitigating Civilian Casualties.*

[6] Associated Press, "U.S. Orders Probe of Fatal Shooting of Italian Officer," NBC News, 8 March 2005, http://www.nbcnews.com/id/7089948/ns/world_news-mideast_n_africa/t/us-orders-probe-fatal-shooting-italian-officer/#.VfgBGrT9pi0.

[7] Though MNF-I did not establish a dedicated tracking cell for civilian casualties as did ISAF, MNF-I Headquarters tracked these casualties for certain periods as visibility of the issue increased. For example, more than 500 civilian casualties resulted from EOF incidents in the first half of 2005.

the conflict, Multinational Force–Iraq (MNF-I) pointed to the strategic importance of EOF and cited the lack of available nonlethal capabilities and inadequate training in their use as key deficiencies.

COIN in Afghanistan

In Afghanistan, President Hamid Karzai made his first public statements regarding CIVCAS in 2005, asking the Coalition to take measures to reduce such casualties. Early initiatives to minimize CIVCAS in Afghanistan, such as the "Karzai 12" rules for approving operations in 2005 and the initial Commander, International Security Assistance Force (COMISAF) tactical directive in 2007, were not successful in reducing high profile incidents.[8] Additional efforts, including redrafting the COMISAF tactical directive in 2008, were made in response to several high-profile, high-CIVCAS incidents; however, another incident in May 2009 in Bala Baluk, Farah Province, where dozens of civilians were killed in U.S. airstrikes, highlighted the lack of progress in effectively addressing civilian casualties.[9]

The Bala Baluk incident served as an impetus for significant efforts to reduce CIVCAS by both ISAF and the United States. Since mid-2009, ISAF leadership clearly and consistently emphasized the importance of reducing CIVCAS, and ISAF modified its policies and procedures to this end. Similarly, concerted efforts on the part of the United States—spearheaded by the U.S. Joint Staff CIVCAS Working Group, led by a three-star general officer—aided efforts to improve predeployment training to better prepare U.S. forces for CIVCAS reduction and mitigation in Afghanistan. This included efforts to provide additional nonlethal and low collateral damage tools, as well as address deficiencies in predeployment training regarding the use of nonlethal tools already available. Collectively, these dedicated efforts bore fruit; because of improved guidance and training, ISAF forces

---

[8] "Karzai 12" refers to a dozen new ROE sent down by Gen McChrystal to reduce CIVCAS. For more on the COMISAF tactical directive, see *Civilian Harm Tracking*.

[9] See "Tactical Directive" (memorandum, ISAF Headquarters, Kabul, Afghanistan, 30 December 2008), http://www.nato.int/isaf/docu/official_texts/Tactical_Directive_090114.pdf; and Carlotta Gall and Taimoor Shah, "Civilian Deaths Imperil Support for Afghan War," *New York Times*, 6 May 2009, http://www.nytimes.com/2009/05/07/world/asia/07afghan.html?_r=0.

adapted how they conducted operations in light of CIVCAS concerns, and ISAF-caused CIVCAS decreased over time. Importantly, analysis of available data suggested these CIVCAS mitigation efforts represented a win-win, with no apparent cost to mission effectiveness or increase in friendly force casualties.

In a review of U.S. operations, cases where the government failed to comply with IHL were rare, and were usually associated with U.S. forces being in close proximity to the population to better protect it from enemy attacks.[10] While IHL cautions military forces to keep a distance from civilians and protected functions, such as medical facilities, tension exists between compliance with IHL in this regard and civilian protection strategies that put U.S. forces between the civilian population and an indiscriminate enemy that often targeted civilians for the purpose of intimidation.

COUNTERTERRORISM OPERATIONS

The United States conducts CT operations to target members of al-Qaeda, the Taliban, ISIL, and affiliated groups. These operations have been conducted both in major theaters of operation (e.g., Iraq and Afghanistan) and in CT campaigns outside declared theaters of operation (e.g., Pakistan and Yemen). The government justifies its CT campaign, in part, based on an imminent threat to U.S. interests and the minimal cost of this approach to civilian lives.

As discussed earlier, U.S. officials have regularly stated that reducing the risk of CIVCAS in its CT operations is a national priority, and that the United States does everything possible to that end. The U.S. government's commitment to minimize civilian harm is laudable. However, its descriptions of civilian casualties for operations in Pakistan and Yemen were significantly lower than every other estimate available, including several open source estimates and a recent UN report.[11] These civilian casualties include those arising from drone strikes, where U.S. officials and some aca-

---

[10] This review was documented in a classified report by Larry Lewis, *IHL Lessons from the Past Decade: Operations in Afghanistan* (Alexandria, VA: CNA, 2015).

[11] Emmerson, *Promotion and Protection of Human Rights*.

demics have described the precision and low collateral damage nature of these strikes with adjectives such as "surgical" and "humane."[12] For CT operations, the striking disparity between the two sets of civilian casualty estimates—and especially unrealistic comments from the U.S. government about zero civilian casualties—injured U.S. credibility and legitimacy in the international community.

The recent case of a U.S. drone strike killing an American hostage in Pakistan in January 2015 illustrates the twin challenges of estimating collateral damage before a strike and characterizing the actual levels of civilian casualties after a strike.[13] In Afghanistan, CIVCAS from airstrikes had the same challenges: only the presence of ground forces, host nation security forces, and communication with the population and international organizations allowed refinements of incomplete BDAs based on air platform observations.[14] As a result, the government's initial estimates, which tended to be too low, could be revised in light of additional evidence.

Improvements have been made over time for U.S. CT operations, according to open source estimates. For example, in Pakistan, available information points to the average number of civilian deaths per drone strike decreasing over time, with the likelihood of CIVCAS (as measured by the rate of operations causing civilian casualties) also decreasing; less than 20 percent of airstrikes caused civilian casualties in 2010, with this rate falling to less than 10 percent for 2013 and dropping to zero for 2014. Overall, U.S. CT operations have become less likely to cause civilian deaths over time.[15]

However, it appears that there is still room for improvement, especially in Yemen. The percentage of U.S. operations in Yemen that caused CIVCAS was significantly higher than those seen for CT operations in

---

[12] Such descriptions from academia include Daniel Byman, "Why Drones Work: The Case for Washington's Weapon of Choice," *Foreign Affairs*, July/August 2013; and Lewis, "Drones."

[13] Adam Entous, Damian Paletta, and Felicia Schwartz, "American, Italian Hostages Killed in CIA Drone Strike in January," *Wall Street Journal*, 23 April 2015, http://www.wsj.com/articles/american-italian-hostages-killed-in-cia-drone-strike-in-january-1429795801.

[14] Sewall and Lewis, *Civilian Harm Tracking*.

[15] Lewis, *Improving Lethal Action*.

Afghanistan conducted by U.S. and international forces. While CIVCAS rates for the two countries are not necessarily directly comparable, operations in Afghanistan illustrate that lower rates can be achieved during CT operations in general. This point is echoed by reduced CIVCAS rates for more recent operations in Pakistan. Also, with regard to CT operations in Afghanistan, ground operations were observed to have a much smaller rate of civilian casualties than airstrikes, illustrating that the type of operation has an effect on the ability to protect civilians from harm during these operations.[16]

---

[16] Ibid.

CHAPTER 22

CONCLUSIONS AND RECOMMENDATIONS

A FINAL WORD

U.S. national security would benefit from an increased consideration of legitimacy within the overall calculus of U.S. national security policy. Lack of security progress in places like Iraq (against ISIL), Nigeria (against Boko Haram), and Mali can be tied to lack of legitimacy. Conversely, the issues in Colombia and the Philippines show that improving legitimacy as part of a full-spectrum security assistance effort can reduce grievances and undermine the factors that fuel conflict. These considerations promote sustainable security, enabling more effective response to threats and possibly even avoiding costly do-overs, such as the current U.S. involvement in Iraq with ISIL.

    Likewise, the U.S. government has improved the humanitarian conduct of its operations during the past decade. Because of early lapses in behavior and a few continuing concerns about the use of force and transparency, however, the international reputation and legitimacy of the nation has been tarnished. Taking steps to learn lessons from these events would enhance the international legitimacy of U.S. actions and demonstrate that our values consistently govern our conduct. These steps would also promote effective security assistance and serve as a model for other countries regarding the responsible use of force, thereby advancing U.S. objectives of promoting the rule of law and accountability within the international community. More important, this also improves the government's ability to further develop partner legitimacy, with expected benefits to security described above. In the spirit of lessons learned, this work lays out specific steps to take for improved legitimacy and national security.

RECOMMENDATIONS

This section concludes with two sets of recommendations. The first set highlights the need for the government to change its approach to working with partner nations, which will improve security assistance and other efforts to promote long-term stability and legitimacy in the future:

A-1. Revisit U.S. security assistance

A-2. Prioritize involvement of U.S. ground forces in security assistance

A-3. Improve promotion of IHL with U.S. partners

In addition, as discussed earlier, the United States showed marked improvement in its own operations with regard to IHL mandates for civilian protection, moving from mixed to more consistent compliance, and even supercompliance in some instances. These practices have been replicated in several theaters and by other nations, for example:

- Civilian harm mitigation tactics developed and employed in Afghanistan are used regularly in current U.S. operations in Syria and Iraq to better protect civilians during airstrikes; and

- U.S. civilian harm mitigation doctrine and handbooks have informed partner nation training in Africa, South America, and Asia.

While commendable, overall progress occurred in the shadow of specific concerns and abuses, such as concerns over civilian casualties in Pakistan and elsewhere. For example, after hostages were killed during a drone strike operation, President Obama remarked: "These aren't abstractions, and we're not cavalier about what we do, and we understand the solemn responsibilities that are given to us."[1] In fact, these instances of civilian harm can serve as a rallying cry for violent terrorist organizations and as a grievance fueling violence; at the same time, they undermined the country's legitimacy and the ability of the U.S. government to influence other nations and actors toward constructive and desirable behavior.[2]

To improve U.S. legitimacy in the future and better enable national security efforts, four recommendations are provided for improving the legitimacy of U.S operations in transparent ways:

B-1. Ensure U.S. institutions are accountable to IHL

B-2. Create organizational focus on IHL compliance and supercompliance

B-3. Perform independent reviews of current operations

B-4. Codify recent U.S. best practices into international norms for the conduct of hostilities

RECOMMENDATION A-1: REVISIT U.S. SECURITY ASSISTANCE

Security assistance was described earlier as having a tactical train-and-equip focus that often neglected the essential tasks of institution-building, as well as changing the mindset and strategy to promote legitimacy and effectiveness. Recently, the U.S. military has increased its focus on building institutions, which represents a move in the right direction. At the same time, more can be done, and this focus should emphasize factors that promote legitimacy as well as effectiveness.

Especially for security assistance where partner nations are facing security threats requiring the use of force, where a full spectrum approach —tactical-level training, institution-building, and providing niche capabilities—with an emphasis on promoting legitimacy would encourage long-

---

[1] Peter Baker and Julie Hirschfeld Davis, "Amid Errors, Obama Publicly Wrestles with Drones' Limits," *New York Times*, 24 April 2015, http://www.nytimes.com/2015/04/25/us/politics/hostage-deaths-show-risk-of-drone-strikes.html?_r=0.

[2] Hassan Abbas, "How Drones Create More Terrorists," *Atlantic*, 23 August 2013, http://www.theatlantic.com/international/archive/2013/08/how-drones-create-more-terrorists/278743/.

term stability and improved national security. These efforts should be resourced adequately, based on a cost-benefit analysis showing that initiatives to avoid or shorten conflict are much less expensive than a potential military response. In the case of the Philippines and Colombia, the model where the embassy served as a command post demonstrated several benefits, including synchronizing U.S. government efforts and building strong relationships and U.S. influence with counterparts. This situation could help the government avoid different and sometimes conflicting agendas that limit forward progress.

RECOMMENDATION A-2: PRIORITIZE U.S. GROUND FORCES' INVOLVEMENT IN SECURITY ASSISTANCE

Recent U.S. policy has resisted the use of troops on the ground, in a sharp contrast to the hundreds of thousands of forces who deployed to Iraq and Afghanistan during the past decade. While the latter level of deployment may not be sustainable or desirable, a policy of zero or minimal U.S. forces on the ground offers strategic costs. The United States should consider dedicating forces on the ground for advise-and-assist roles in security assistance cases where the host nation is actively involved in armed conflict. Without ground forces, the government abdicates its ability to influence and shape the operational approach, which includes making military efforts effective to reduce prolonged conflict and helping the host nation improve its legitimacy. This also includes using American influence to reduce abuses of power and encourage approaches that resolve grievances fundamental to the conflict.

Most notably, this kind of interaction does not require U.S. forces to take direct combat roles. In both the Philippines and Colombia, U.S. policy prevented military forces from assuming combat roles, yet they had a significant positive effect on host nation forces through tactical advising, reinforcing a will to fight, and imparting a population-centric ethos. By foregoing this potential role for U.S. forces, the government loses a chance to positively influence other nations and promote their legitimacy. Lack of U.S. involvement also provides opportunities for other nations that do not share these same values to influence the host nation population and possibly to perpetuate the conflict. An immediate application of this recommendation would be to expand the number of U.S. advisors on the ground in Iraq, using them similarly to how Special Forces were employed in the Philippines, to advise and assist Iraqi Security Forces and Kurdish Peshmerga forces at the tactical level as they strive to counter ISIL.[3] Partnering U.S. forces with those on the ground would involve changes to organization, equipment, and training/education to match the requirements for this role.

RECOMMENDATION A-3: IMPROVE PROMOTION OF IHL WITH U.S. PARTNERS

The United States exerts considerable effort to provide security assistance to build security force capacity and proficiency in partner nations. Under current U.S. law, humanitarian considerations remain paramount in national decisions to conduct security assistance; patterns of IHL violations can halt such assistance to units and even entire nations due to the Leahy Law.[4] While IHL compliance affects decisions regarding whether to provide assistance, however, the U.S. approach to promoting IHL compliance remains incomplete. The current U.S. approach—chiefly providing education and training on IHL—is based on the tenet that IHL violations stem from ignorance of the requirements of international law. In fact, this

---

[3] "ISIL and Peshmerga Forces Battle for Ground Near Kirkuk," Al Jazeera, 6 July 2015, http://www.aljazeera.com/news/2015/07/isil-peshmerga-forces-battle-ground-kirkuk-150706042712437.html.

[4] Written in 1996 to control U.S. military aid to Colombia and named for Senator Patrick J. Leahy, the Leahy amendment (22 U.S. Code § 2304 - Human rights and security assistance) seeks to promote host nation accountability and to motivate nations to address patterns of gross violations. However, some argue that the Leahy Law should not be applied in cases where nations face exigent threats. Eric Schmitt, "Military Says Law Barring U.S. Aid to Rights Violators Hurts Training Mission," *New York Times*, 20 June 2013, http://www.nytimes.com/2013/06/21/us/politics/military-says-law-barring-us-aid-to-rights-violators-hurts-training-mission.html.

is often not the case.[5] For example, IHL violations are often committed by forces who are well aware of international laws but decide it is more operationally expedient to break the rules. However, case studies of events in Colombia and the Philippines show operational benefits from reducing IHL violations and more effectively protecting the population.[6] To more effectively promote compliance in partner nations, the U.S. and international community needs to move beyond their current approach by making the case to partner nations that operational expediency is not a sufficient justification for violating IHL and can, in fact, be counterproductive in the long term.[7] The recommendations for U.S. actions below are important to the success of this effort; if compliance is the behavior that the U.S. government wants to promote in others, it must consistently model legitimacy in its own actions.

There are also cases where partners seem to follow IHL, but their operations still result in high civilian tolls or other negative humanitarian impacts. This situation hurts the partner nation, creating grievances and exacerbating security challenges, and injures U.S. legitimacy. In those cases, the United States should work to help partner nations consider creative ways to reduce civilian casualties from their operations. The government could also consider instituting proactive training and education to enhance partner capabilities for reducing civilian casualties. These initiatives could be tied to other elements of security assistance, such as the DOD's Foreign Military Sales (FMS) program.[8]

---

[5] Daniel Muñoz-Rojas and Jean-Jacques Frésard, *The Roots of Behaviour in War: Understanding and Preventing IHL Violations* (New York: International Committee of the Red Cross, 2004), https://www.icrc.org/eng/assets/files/other/irrc_853_fd_fresard_eng.pdf.

[6] MajGen Geoffrey Lambert, USA (Ret), Dr. Larry Lewis, and Dr. Sarah Sewall, *The Salience of Civilian Casualties and the Indirect Approach: Operation Enduring Freedom-Philippines* (Washington, DC: Joint Forces Command, 2012); and *Learning from Colombia*.

[7] This can serve as motivation for nations to better operationalize IHL by incorporating it into policy, doctrine, senior leader education, and accountability processes.

[8] For more information, see Defense Security Cooperation Agency, FMS program at http://www.dsca.mil/programs/foreign-military-sales-fms.

RECOMMENDATION B-1. ENSURE U.S. INSTITUTIONS ARE ACCOUNTABLE TO IHL

One safeguard for the future is to make sure institutions involved in armed conflict activities (e.g., the use of force) are held accountable to IHL. For example, the U.S. military is legally obligated to comply with IHL; it is built into the U.S. Uniform Code of Military Justice (UCMJ), and forces can be criminally charged for IHL violations when necessary.[9] While such accountability measures do not always prevent abuse, they avoid institutional impunity and serve as a deterrent to make it less likely that others will follow the same example.

Of course, taking steps to ensure IHL compliance is not simply a legal consideration. For the U.S. military, ethics are a fundamental part of the profession of armed service. For example, Chairman of the Joint Chiefs of Staff General Dempsey provided the following guidance to all members of the military:

> Our oath demands each of us display moral courage and always do what is right, regardless of the cost. . . . Commitment to the rule of law is integral to our values which provide the moral and ethical fabric of our profession.[10]

In addition, the U.S. military as an institution incorporates measures meant to ensure compliance with IHL. For example, IHL principles are built into the military's policy, doctrine, training, and accountability processes. When civilian casualties were suspected in Afghanistan and in recent operations in Iraq and Syria, command-directed investigations were conducted routinely. These investigations were used as tools for learning, but they also looked for possible violations of ROE or any other IHL considerations.

---

[9] The UCMJ (64 Stat. 109, 10 U. S. C. Chapter 47) serves as the foundation of U.S. military law. UCMJ applies to all members of the uniformed Services: the Air Force, Army, Coast Guard, Marine Corps, Navy, National Oceanic and Atmospheric Administration Commissioned Corps, and Public Health Service Commissioned Corps. For more information, see http://www.ucmj.us.

[10] Martin E. Dempsey, *America's Military: A Profession of Arms* (white paper, Joint Chiefs of Staff, Washington, DC, 2012), 3.

While such violations were rare, this practice promoted a culture of military accountability to IHL. These additional measures—doctrine, training, policy, and a culture of accountability—should also be developed in other U.S. organizations that conduct armed conflict activities (e.g., the use of lethal force) in the future.

RECOMMENDATION B-2. CREATE AN ORGANIZATIONAL FOCUS ON IHL

During the past decade of operations, one best practice stands out when resolving humanitarian concerns: when the U.S. military established an organizational focus on that issue. For example, when CIVCAS became a strategic issue for operations in Afghanistan, two organizations were created to reduce CIVCAS and improve civilian harm mitigation: the Civilian Casualty Mitigation Team within ISAF headquarters in Kabul and the Joint Staff Civilian Casualties Working Group in the Pentagon. Both organizations yielded tangible improvements.

Note, however, that these two efforts represent ad hoc measures intended to respond quickly to immediate problems within specific operations. In contrast, currently no institutional office or position exists, such as a deputy assistant secretary of defense, to address the conduct of hostilities. Thus, while the ad hoc measures were successful in the short term for improving operations, their ad hoc nature leaves an institutional gap in DOD for policies and practices regarding the conduct of operations with respect to IHL.

The lack of an institutional focus for the conduct of operations, within DOD or elsewhere, complicates the process of transferring key lessons from one operation to another.[11] This means institutional organizations cannot inform current operations in Iraq and Syria. There is also no such institutional focus for CT operations in Yemen and Pakistan, which include activities beyond those by the U.S. military. When faced with this lack of focus and commensurate expertise, the U.S. government may turn to nonexperts

---

[11] "One of the enduring challenges of military operations [is] the difficulty of taking hard-fought lessons from one theater and applying them to others." Alexander Powell et al., *Summary Report: U.S.–UK Integration in Helmand* (Arlington, VA: CNA, February 2016).

for its policy and practice, which can have disastrous results, as evidenced by the inappropriate use of the Joint Personnel Recovery Agency (JPRA) as advisors for detainee interrogations after 9/11.[12]

RECOMMENDATION B-3. PERFORM INDEPENDENT REVIEWS OF CURRENT OPERATIONS

For military CT actions, operational assessments were often done by the organization conducting the operation, without the benefit of independent reviews. Thus, congressional oversight of those operations generally relied on the organizations to grade their own papers.

Independent reviews bring value to an organization for several reasons. Besides being helpful when the group faces a potential conflict of interest, an objective external review may identify blind spots and misconceptions. Illustrating the value of independent reviews, the U.S. military and international forces operating in Afghanistan benefitted from multiple independent reviews to improve their ability to reduce civilian harm and increase their legitimacy.[13] These examples illustrate in-stride reviews that informed current as well as future operations and led to improved guidance, tactics, and training. They also reveal areas of misunderstanding regarding how CIVCAS occurred; in these cases, existing guidance and tactics were ineffective because they failed to address actual root causes. Independent reviews featuring an evidence-based approach enabled tailored solutions that were seen as win-win opportunities—mission success improved while civilian casualties decreased. Such a review process should be standard practice in U.S. operations, starting with examining the issue of civilian harm in CT operations in Yemen, Pakistan, Syria, and Iraq.

Note that international expectations for transparency and accountability can collide with the inherent nature of covert operations. However, "need to know" can at times be used as a cloak to avoid accountability. Importantly, objective and independent reviews can still be conducted at classified levels.

---

[12] In fact, it has already done so with regard to lessons from Israel on the conduct of operations, as discussed later in this work.

[13] A complete list of these reviews is included in Lewis, *Reducing and Mitigating Civilian Casualties*.

In fact, the U.S. military has a history of such reviews, requesting them both for civilian harm mitigation in Afghanistan (as described above) and for overall lessons regarding IHL compliance from Afghanistan operations.[14] Establishing such independent reviews as a standard practice for future operations, covert or not, could identify and mitigate patterns of IHL violations earlier; for example, the reviews could mitigate the need for congressional after action reports, which provide transparency only after the fact, with no ability to shape ongoing operations or shape public perceptions of U.S. legitimacy.

RECOMMENDATION B-4. CODIFY RECENT U.S. BEST PRACTICES IN INTERNATIONAL NORMS FOR THE CONDUCT OF HOSTILITIES

As described above, the United States has made distinct progress in compliance—and indeed supercompliance—with IHL over the past decade, though with a few exceptions. Codifying these best practices would help the government be more consistent in sustaining this progress in future operations.

A similar opportunity exists for recent U.S. progress in the conduct of hostilities. A Copenhagen-like effort focused on best practices and principles would be beneficial for addressing the complexities of modern warfare, including drones use, targeting individuals, and effectively dealing with modern insurgent tactics that endanger noncombatant civilians.[15] Such an effort should include the recent progress made in Afghanistan, Iraq, and Syria regarding protection of civilians, as well as similar progress made by the situations in the Philippines and Colombia.

---

[14] Larry Lewis, *IHL Lessons from the Past Decade: Operations in Afghanistan* (Arlington, VA: CNA, 2015).

[15] Refers to the Copenhagen Process, which was launched by the Danish Government in 2007 to address a range of practical and legal challenges to nations and organizations involved in international military operations. For our purposes, the report published by the group also applies to IHL, armed conflict, and peace operations. See *The Copenhagen Process: Principles and Guidelines* (Copenhagen, Denmark: Ministry of Foreign Affairs, 2007), http://um.dk/en/politics-and-diplomacy/copenhagen-process-on-the-handling-of-detainees-in-international-military-operations/~/media/368C4DCA08F94873BF1989BFF1A69158.pdf.

This effort would complement current U.S. initiatives regarding technology proliferation. For example, the government recently revised its export policy regarding drones for military use.[16] This policy includes an assessment of the receiving country's intent for use of these platforms and an evaluation that this use conforms to IHL.[17] However, the past decade has shown that tragic events can happen even when a force is complying with IHL, which drove the United States to develop best practices above and beyond IHL to improve civilian protection within its legal use of force. Developing the conduct-of-hostilities equivalent to the Copenhagen Process would ideally advance international practice as well as advance U.S. interests, such as the concerns evident in the recent U.S. drone export policy.

CLOSING

The United States has been generally effective in attaining the short-term stability of host nation governments after dedicating adequate resources; however, the equally critical goal of reaching longer-term stability and legitimacy has been more elusive. U.S. forces are back in Iraq yet again, this time to deal with ISIL; likewise, security assistance in Africa has not stopped repeated uprisings from the Tuareg in Mali or continuing problems from Boko Haram in Nigeria.[18] As seen in these examples, failure to promote long-term stability creates opportunities for potential threats to fester; for example, armed groups fill a security and governance void and become stronger until they jeopardize the host-nation government and pose a real threat to U.S. interests. By improving the legitimacy of the United States and its partners, the recommendations above provide a path for improving long-term national security and international stability.

---

[16] Andrea Shalal and Emily Stephenson, "U.S. Establishes Policy for Exports of Armed Drones," Reuters, 18 February 2015, http://www.reuters.com/article/2015/02/18/us-usa-drones-exports-idUSKBN0LL21720150218.

[17] "U.S. Export Policy for Military Unmanned Aerial Systems" (fact sheet, U.S. Department of State, 17 February 2015), http://www.state.gov/r/pa/prs/ps/2015/02/237541.htm.

[18] For more on these issues, see Anthony N. Celso and Robert Nalbandov, ed., *The Crisis of the African State: Globalization, Tribalism, and Jihadism in the Twenty-First Century* (Quantico, VA: Marine Corps University Press, 2016).

APPENDIX A

TITLE 10, TITLE 50, AND OVERSIGHT

As is the case with international law, U.S. domestic law regulates when and how the United States may use force.[1] Both Congress and the executive branch play important roles in these processes, and this appendix concentrates on aspects of those roles that are embodied in Titles 10 and 50 of the U.S. Code, with a focus on oversight. This appendix aims to clarify the interplay of the two titles in drone strike (and other CT) operations and to provide background on the military preference policy. In doing this, it begins a discussion of covert actions that is continued in the next appendix.

Titles 10 and 50 are often misconstrued in the debate over drone strikes, where "Title 10" is used as shorthand for the military and its actions, and "Title 50" for the intelligence agencies and their actions.[2] In fact, the dis-

---

[1] Several circumstances exist in which the use of force is permitted. The president has an implicit duty under Article II of the U.S. Constitution as commander in chief to defend the nation. In addition, Congress may declare war or otherwise authorize the use of force through legislation. Finally, the War Powers Resolution allows the president to authorize the use of force based on immediate need and without congressional approval for 60 days; after which time, if Congress has not acted, the president has another 30 days to withdraw U.S. forces.

[2] For example, Andru E. Wall, "Demystifying the Title 10-Title 50 Debate: Distinguishing Military Operations, Intelligence Activities, and Covert Action," Harvard Law School *National Security Journal 3*, no. 1 (December 2011), http://harvardnsj.org/2011/12/demystifying-the-title-10-title-50-debate-distinguishing-military-operations-intelligence-activities-covert-action/; and Josh Kuyers, "CIA or DOD: Clarifying the Legal Framework Applicable to the Drone Authority Debate," American University Washington College of Law *National Security Law Brief*, 4 April 2013.

tinction is not so clear-cut—the military can act under Title 50, as is detailed below—but oversight mechanisms constitute an important difference.

Title 10 provides for the structure and general powers of the armed forces, as well as the oversight mechanisms for most military activities. In particular, actions such as traditional military activities (TMA) fall under Title 10, and as such are overseen by the Senate and House Armed Services Committees. TMA is not defined in the law, although Congress has provided some nonstatutory guidance (see Appendix B).

Title 50 governs—among other things—intelligence collection, covert actions (as defined in the title), and the oversight of these activities. It also governs the structure and some functions of the CIA and aspects of the intelligence community more broadly.

Title 50 defines a covert action as ". . . an activity or activities of the United States Government to influence political, economic, or military conditions abroad, where it is intended that the role of the United States Government will not be apparent or acknowledged publicly . . ." while exempting TMA, traditional diplomatic activity, intelligence collection, and several other types of activities.[3] Title 50 specifies that the president may authorize covert actions by means of written findings. Although the CIA is considered the traditional agency for carrying out covert actions, Title 50 makes reference to the possibility of other departments or agencies carrying out the actions so, in particular, DOD may carry out covert actions (as defined in Title 50) such as covert drone strikes, and thus act under Title 50.[4] Whether DOD carries out "Title 50 covert" actions in practice is a separate issue not addressed here, although this would have implications for the military preference policy.

Title 50 stipulates congressional oversight for covert actions, requiring "the Director of National Intelligence and the heads of all departments, agencies, and entities of the United States Government involved in a covert

---

[3] Title 50, U.S. Code, § 3093 (e). This definition differs from DOD's doctrinal definition of a covert action, which is discussed further in Appendix B.

[4] See *United States Intelligence Activities*, Executive Order No. 12333, 46 Federal Register 59941, 3 Code of Federal Regulations (1981), Part 1, 1.8(e).

action . . . [to] keep the congressional intelligence committees fully and currently informed of all covert action," although in exceptional circumstances, activities can be temporarily revealed only to the so-called "Gang of Eight" (i.e., the majority and minority leaders of the Senate, the Speaker and minority leader of the House, and the chairmen and ranking minority members of the House and Senate Intelligence Committees).[5] Senate Intelligence Committee Chairwoman Feinstein has put forth that her committee "receive[s] notification with key details shortly after every [drone] strike, and . . . [holds] monthly in-depth oversight meetings" that look rigorously at the drone program.[6] Issues surrounding covert actions are discussed at greater length in Appendix B.

One significant issue of interest here is that the oversight and accountability mechanisms in Title 50 are triggered by reasonably subjective criteria for the DOD. Indeed, as noted above, TMA is not defined in U.S. law, yet any activity classified as TMA is exempt from the requirements for covert activities of a presidential finding and oversight by the House Intelligence Committee or the Senate Select Committee on Intelligence. Similarly, there can be overlap between intelligence collection (which falls under Title 50 and is overseen by the intelligence committees) and TMA, such as operational preparation of the environment (OPE), which can include, for example, significant intelligence collection from a site in advance of an attack but falls under Title 10. DOD's interpretations of TMA and OPE have been described as "broad."[7]

As it stands, this ambiguity leaves the DOD open to the perception that it is (or has the potential of) circumventing Title 50 oversight mechanisms by classifying its activities as TMA.[8] Indeed, broad interpretations of these

---

[5] Title 50, U.S. Code, § 3093 (b).

[6] Senator Dianne Feinstein, "Letters: Senator Feinstein on Drone Strikes," *Los Angeles Times*, 17 May 2012, http://articles.latimes.com/2012/may/17/opinion/la-le-0517-thursday-feinstein-drones-20120517.

[7] Eric Schmitt, "Clash Foreseen Between CIA and Pentagon," *New York Times*, 10 May 2006, http://www.nytimes.com/2006/05/10/washington/10cambone.html?pagewanted=all.

[8] See, for example, Erwin, *Covert Action*.

missions would circumvent the significant requirement of a presidential finding, but this perception also may presuppose that Title 10 oversight by the House and Senate Armed Services Committees is somehow preferable to DOD in certain cases over Title 50 oversight by the intelligence committees. In fact, pundits have made various further claims that the oversight of the intelligence committees or armed services committees is qualitatively superior to that of the other. The basis for either assertion has not been satisfactorily clarified in any writings to date. Indeed, the congressional oversight process with respect to such issues as drone strikes is poorly understood by the American public for, although public hearings can get substantial media coverage, the reporting requirements and other means of oversight employed by Congress and its committees are not well explained by Congress. Comparing the oversight of these committees would be a worthwhile venture, but would be best served as part of a more thorough public explanation of the oversight process. In particular, such a comparison would have important ramifications for the military preference policy.

In any case, if the government were to describe how it defines TMA and OPE and how it distinguishes them from Title 50 activities, that would increase the transparency (and perhaps the consistency) of the oversight process, although this might cost DOD some of the flexibility provided by the existing ambiguity.

One relevant practice that ties into all of these issues is that of one government agency acting in a supporting role for an operation led by another agency. Thus, in addition to DOD acting under Title 10 or Title 50 and OGAs acting under Title 50, DOD may provide forces in support of an operation under the direction and authority of an OGA, and similarly an OGA may act in support of a DOD operation. These practices may be practical and aid tactical and operational effectiveness, however, as with the lack of a definition for TMA, they hinder transparency and could leave the perception that the agencies involved are skirting some oversight mechanisms. This topic is discussed further in the chapter on legitimacy issues.

APPENDIX B

COVERT ACTIONS

This appendix provides a more in-depth discussion of covert actions within U.S. domestic law. It aims to articulate and clarify some of the issues related to covert actions, and presents further considerations with regard to the military preference policy, particularly when that policy includes a preference for strikes to be carried out under Title 10. If unqualified, "covert" will be taken to mean an action by a government that, at the time the action is carried out, is not intended to be acknowledged by the government.

Covert actions have a long history of being carried out by nations for military, national security, and diplomatic ends. Espionage serves as a primary example, and is considered legal under international law as a form of self-defense. Espionage, as a general practice, is not particularly controversial, perhaps because there is a tacit understanding that "everyone does it" and it is not understood to be particularly violent. Covert military (or paramilitary) actions do not share these comforts, although they are legally justified as being derived from customary international law.

It has been widely reported that the United States has carried out covert drone strikes in Pakistan, with the conventional wisdom that they were performed with secrecy so as to allow the Pakistani government the ability to plausibly deny that the strikes are taking place (although the Pakistani government has in recent years acknowledged them via condemnations).[1] Furthermore, various media outlets have asserted that the CIA must perform these operations, with the implication that DOD does not carry out covert drone strikes, although Appendix A discusses how DOD is not legally barred from conducting such activities.[2]

Noncommittal and conflicting statements have come out of high levels of the U.S. government on the subject of the U.S. military carrying out covert actions. In his 2007 confirmation hearing, Undersecretary of Defense for Intelligence and current Director of National Intelligence Lieutenant General James R. Clapper Jr. testified that Title 50 covert activities "are *normally* not conducted . . . by uniformed military forces," tacitly acknowledging that DOD forces conduct covert actions.[3] However, in his written testimony for the same hearing, Clapper said it was his understanding that "military forces are not conducting 'covert action'," but are limiting themselves to clandestine action.[4] He went on to explain that he had been referring to the passive/active distinction given below.

The notion that only the CIA may perform covert actions or that DOD is barred from performing them is indeed pervasive, not only within the media but also within the defense community and the government.[5] As noted above, DOD not only appears to be permitted to carry out covert action under U.S. law, but its own doctrine makes reference to conducting such actions. This provides evidence that U.S. policy with respect to covert actions is unclear, both for the general public and for those within the defense community.

---

[1] For more on the Pakistani government's stance, see Agence France-Presse, "High Cost of Technology: Pakistan Condemns North Waziristan Drone Strike," *Express Tribune*, 26 December 2013, http://tribune.com.pk/story/650880/pakistan-condemns-north-waziristan-drone-strike/.

[2] This assertion can be seen in, for example, Micah Zenko, *Transferring CIA Drone Strikes to the Pentagon*, Policy Innovation Memorandum No. 31 (New York: Council on Foreign Relations, April 2013), http://www.cfr.org/drones/transferring-cia-drone-strikes-pentagon/p30434.

[3] Erwin, *Covert Action*. Emphasis added.

[4] Ibid.

[5] See, for example, Senator Clarence W. Nelson's more recent questioning of Gen Clapper. Clarence W. Nelson, "Testimony on Current and Future Worldwide Threats to the National Security of the United States" (testimony, U.S. Senate, Committee on Armed Services, 11 February 2014), 5–8, http://www.armed-services.senate.gov/imo/media/doc/14-07%20-%202-11-14.pdf.

Contributing to the confusion is the lack of consensus on the meaning of the word *covert*. *Covert action* is defined in Title 50; a *covert operation* is defined in military doctrine (the definitions are similar but not identical); and the term *covert* is used in even different ways colloquially and in General Clapper's Senate testimony, where he ascribes to the word an active/passive meaning. These usages are given in Appendix table 1, as is the doctrinal definition of *clandestine*.

The table shows that the "passive/active" alternate DOD characterization by General Clapper is notably different from the other usages (and perhaps would be more straightforward for operators and lawyers to work with).

Appendix table 1. Definitions of covert

Usage	Definition	References
U.S. law (Title 50)	. . . [T]he term **covert action** means an activity or activities of the United States Government to influence political, economic, or military conditions abroad, where it is intended that the role of the United States Government will not be apparent or acknowledged publicly, but does not include 1. Activities the primary purpose of which is to acquire intelligence, traditional counterintelligence activities, traditional activities to improve or maintain the operational security of United States Government programs, or administrative activities; 2. Traditional diplomatic or military activities or routine support to such activities; 3. Traditional law enforcement activities conducted by United States Government law enforcement agencies or routine support to such activities; or activities to provide routine support to the overt activities (other than activities described in paragraph (1), (2), or (3)) of other United States Government agencies abroad. [Emphasis added.]	Title 50, U.S. Code, Section 3093

DOD doctrine	**Definition of covert operation**: an operation that is so planned and executed as to conceal the identity of or permit plausible denial by the sponsor. Note also the **definition of a clandestine operation**: an operation sponsored or conducted by governmental departments or agencies in such a way as to assure secrecy or concealment. A clandestine operation differs from a covert operation in that emphasis is placed on concealment of the operation rather than on concealment of the identity of the sponsor. In special operations, an activity may be both covert and clandestine and may focus equally on operational considerations and intelligence-related activities.	DOD Joint Publications[6]
Alternate DOD characterization	Although testifying that the term *clandestine activities* is not defined by statute, [General Clapper] characterized such activity as consisting of those actions that are conducted in secret, but which constitute "**passive**" intelligence information gathering. By contrast, covert action, he suggested, is "**active**," in that its aim is to elicit change in the political, economic, military, or diplomatic behavior of a target. [Emphasis added.]	Clapper testimony[7]
Colloquial usage	Any secret action	Standard dictionaries[8]

The colloquial usage of the word "covert" lacks the subtleties of how the term is used in law and doctrine. In particular, note that the DOD and Title 50 definitions allow for the disclosure of a covert action after the fact, as long as there is the intent for nonattribution at the time the action is done.[9] This allowed, for example, then-CIA Director Leon Panetta to

---

[6] *Department of Defense Dictionary of Military;* and *Joint Special Operations.*

[7] Erwin, *Covert Action.*

[8] For example, Merriam-Webster, http://www.merriam-webster.com/dictionary/covert.

[9] Journalists and scholars sometimes indicate that a covert action *may not* be acknowledged by the government after the fact. See, for example, Erwin, *Covert Action*; and Safire, "Covert Operation, or Clandestine?." This does not appear to be supported by either Title 50 or doctrine. Moreover, material is often classified for a specified window of time, such as several decades.

describe the raid that killed Osama bin Laden as a "covert operation"[10] when he and President Obama presented it to the American public after the attack.[11] Conversely, an activity would not be considered covert if it is intended to be acknowledged—even though there are no statutory limits on when such acknowledgement must take place. This means that, at least in theory, an operation could go unacknowledged by the U.S. government for years after it took place, and it could still be considered noncovert and need not fall under Title 50. Thus, if one hypothetically wanted to "game the system," one would find a weak line dividing the covert from the noncovert.

In the remainder of this appendix, "Title 50 covert" refers to being covert under Title 50, and "doctrinal covert" refers to being covert under DOD doctrine. Recall that covert, if unqualified, refers here to a government action that is intended to be unacknowledged.

The "Title 50 covert" and "doctrinal covert" definitions are subtly distinct. Indeed, "doctrinal covert" actions are DOD actions, whereas the "Title

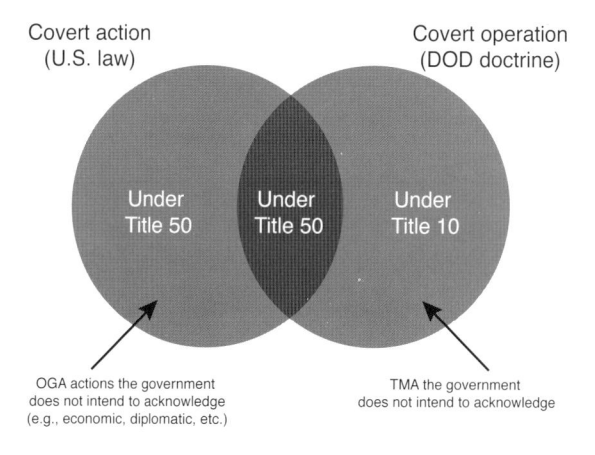

Appendix figure 1. Relationship between legal and doctrinal definitions of covert

Covert action (U.S. law) — Under Title 50 — OGA actions the government does not intend to acknowledge (e.g., economic, diplomatic, etc.)

Under Title 50

Covert operation (DOD doctrine) — Under Title 10 — TMA the government does not intend to acknowledge

---

[10] "CIA Chief Panetta."

[11] Ibid.; and Helene Cooper, "Obama Announces Killing of Osama bin Laden," *The Lede* (blog), *New York Times*, 1 May 2011.

50 covert" actions can be broader, potentially including political, economic, diplomatic, and other activities as well. In addition, the definition of "doctrinal covert" has none of the exemptions contained in the definition of "Title 50 covert," such as the exemption for TMA. This means that an operation could theoretically be considered TMA under U.S. law—and therefore not "Title 50 covert"—but simultaneously be "doctrinal covert" (see Appendix figure 1). Thus the military can carry out covert—i.e., unacknowledged—actions under Title 50 and under Title 10, based on whether the legal or doctrinal definition of covert is used.

As noted above, the definition of a "Title 50 covert" action exempts TMA, but does not define TMA. A former acting CIA general council noted that coming up with a statutory definition of TMA has been "exceedingly difficult."[12] Nonetheless, in the early 1990s, the House of Representatives expressed the following intent for the term's meaning:

> It is the intent . . . that "traditional military activities" include activities by military personnel under the direction and control of a United States military commander (whether or not the U.S. sponsorship of such activities is apparent or later to be acknowledged) preceding and related to hostilities which are either anticipated (meaning approval has been given by the National Command Authorities for the activities and for operational planning for hostilities) to involve U.S. military forces, or where such hostilities involving United States military forces are ongoing, and, where the fact of the U.S. role in the overall operation is apparent or to be acknowledged publicly. [This is intended] . . . to draw a line between activities that are and are not under the direction and control of the military commander. Activities that are not under the direction

---

[12] Matthew C. Dahl, "Event Summary: The bin Laden Operation–The Legal Framework" (American Bar Association, Committee on Law National Security, 2013), http://www.americanbar.org/content/dam/aba/administrative/litigation/materials/sac_2012/50-7_nat_sec_bin_laden_operation.authcheckdam.pdf.

and control of a military commander should not be considered as "traditional military activities."[13]

Covert actions can be controversial. While there is nothing intrinsically illegal about them—from both the domestic and international legal perspectives—the body of international law governing them is quite thin. Moreover, such a level of secrecy might indicate the potential for violating legal principles, such as state sovereignty and IHRL, and the public is often uncomfortable with nations executing such secret actions.

Nonetheless, covert action can be a highly useful tool for nations in that, if it is successful, it might allow a nation to achieve a mission without negative diplomatic or political consequences. Moreover, the denial of a covert action can be a useful diplomatic tool even if the action is known to the other country. One legal scholar pointed out that

> It is less provocative and less disruptive to diplomatic relations not to acknowledge an operation even if the country adversely affected by it is well aware of one's involvement. The target country, either in the interests of good relations or because it cannot effectively prevent it, may ignore the covert action; it is much harder for it to do so if the government conducting it publicly acknowledges what it is doing.[14]

If revealed, however, a covert action has the potential to inflame tensions between the nation executing the action and the nation against which the action took place (or potentially the international community as a whole); this was seen between the United States and Pakistan in the aftermath of the Osama bin Laden raid.

A component of the military preference might be that drone strike operations be carried out under Title 10—in other words, that the strikes not

---

[13] David K. McCurdy, *Conference Report on H. R. 1455* (Washington, DC: U.S. House of Representatives, 25 July 1991), H5905, http://www.fas.org/irp/congress/1991_cr/h910725-ia.htm.

[14] Goldsmith, "Fire When Ready."

be carried out covertly as defined by Title 50. The alternatives for the military carrying out a "Title 50 covert" drone strike are to:

- not act;
- carry out the strike and acknowledge it;
- achieve the same ends through some other course of action;
- carry out the strike in a clandestine manner; or
- carry out the strike as a doctrinal-covert TMA.

The first option—not acting—carries with it the risk of not achieving the mission, or in this situation, leaving an individual deemed a threat to U.S. security untouched. The second option—carrying out a strike and acknowledging it—risks heightening international tensions and perhaps invoking military conflict, even taking for granted that the United States is implementing a sound framework to justify the strikes it carries out. Note that the risk of tensions or conflict would be more severe if even a single acknowledged strike were considered improper or in violation of international law.

The third option—achieving the same ends through another course of action—might carry a high monetary cost or risk to U.S. forces, if such an option would even be possible. For example, alternatives to a covert drone strike could include inserting a SOF team into a hostile area to capture or kill the targeted individual, or launching a full-scale assault into the area to capture or kill multiple targeted individuals. However, such an operation might not be any more palatable than a drone strike. Note also that, if any targets are captured, the U.S. government must then hold and try them either domestically or abroad—a task that can be difficult, as seen during the recent conflicts in Iraq and Afghanistan.

The fourth option raises an important question: to what extent does a covert drone strike differ from a clandestine one? Furthermore, to what extent can a drone strike be covert or clandestine at all? Starting with the second question, clearly evidence from a drone strike—which has reportedly even included weapon debris with U.S. military markings—cannot be hidden from the locals in the area of the attack.[15] Thus the government may not be able to carry out truly clandestine drone strikes. Furthermore, while the United States can always refuse to acknowledge a strike—if even just for some amount of diplomatic cover, as described in the quotation above—given that few other countries are known to operate armed drones, the United States may not be able to carry out drone strikes with much true plausible deniability.

However, a strike can potentially be unknown to the international community or the broader public in the country where the strike took place (i.e., clandestine on a "large scale") if it is done in a very remote location, or if it is not publicized by the government of that country, the media, or social media. Given the current state of global drone operations noted above, a strike could probably only remain unacknowledged to a broader community if it was unknown to that community. Note that this clandestine prerequisite for covert action will only hold as long as other countries largely refrain from using armed drones.

Returning to the first question, it is not clear that any practical differences exist between military drone strikes that are clandestine and military strikes that are covert (as well as clandestine), aside from internal U.S. government processes, such as the differing oversight and approval requirements for "Title 50 covert" actions; military and OGA operators and policymakers would be better positioned to speak to this topic. Again, however, this situation will remain only until other countries begin or increase armed drone operations.

This discussion highlights the fact that the full implications of the military preference, if it were to include a Title 10 preference, may not be realized for a number of years, if and when the use of armed drones becomes

---

[15] For more on debris with U.S. markings, see Greg Miller, "Obama's New Drone Policy Leaves Room for CIA Role," *Washington Post*, 25 May 2013, https://www.washingtonpost.com/world/national-security/obamas-new-drone-policy-has-cause-for-concern/2013/05/25/0daad8be-c480-11e2-914f-a7aba60512a7_story.html.

more prevalent throughout the world. At that point, nonclandestine covert drone strikes would be more viable, so the military preference policy with a Title 10 preference would be significantly more restrictive than the policy would be without a Title 10 preference.

The final option is to accomplish the mission that would have involved a "Title 50 covert" drone strike with a "doctrinal covert" strike that is classified as TMA, if possible. Military operators and lawyers could speak to what extent this would represent a tactical or operational restriction, based on their guidelines for what constitutes TMA and what types of situations in which drone strikes are carried out could not be classified as such.

In the current state of the world (as opposed to when nonclandestine covert strikes may be possible), if indeed a restriction to TMA or clandestine operations instead of "Title 50 covert" ones makes little operational difference, then there is less distinction between the military preference policy with and without a Title 10 preference. Otherwise, if the military preference were implemented with a Title 10 preference, the United States would have to navigate the alternatives to "Title 50 covert" drone strikes, and potentially incur the risk of a significant number of the negative consequences outlined.

INDEX

Bold indicates page with illustration.

Abu Ghraib, Iraq, 189
Afghanistan, 3–4, 9, 11–13, 15–18, 21–23, 27–28, 31–33, 38, 41, 75, 84, 105, 109–10, 112, 119–20, **121**, 122, 124–25, 127, 140–49, 151–54, 157–61, 166, 171, 184, 186–87, 189–96, 198, 200, 203–6, 220
after action report (AAR), 21, **22**, 125, 141, 164, 206
aircraft, xi, 1, 191
    Bell OH-58 Kiowa Warrior helicopter, 142, 144
    Lockheed AC-130 gunship, 141
    McDonnell Douglas F/A-18 Hornet jetfighter, 147
al-Awlaki, Anwar, 53, 56
al-Qaeda in the Arabian Peninsula (AQAP), 53fn15, 54

al-Qaeda in the Islamic Maghreb (AQIM), 178
air-to-surface missile (ASM), 21fn2
Amnesty International, 158, 181fn16
analysis framework, 37–101, 106, 116, 120, 126–27, 164, 220
armed conflict, xii, 2, 6fn1, 67–68, 72–73, 77–78, 80, 97, 99, 172–73, 189, 200, 203–4, 206fn15
    conduct of hostilities, 189–90, 199, 204–7
    detention, 38, 189, 190fn2
    See Law of Armed Conflict (LOAC)
Armed Services Committees, House and Senate, 55, 61, 80, 96, 210, 212
assessment, 13, 15–19, **22**, 23, 27–31, 59, 118, 153–54, 192, 207
    See battle damage assessment

223

Authorization for the Use of Military Force (AUMF), 39, 67fn4, 156

Bala Baluk, Farah Province, Afghanistan, 193
battle damage assessment (BDA), 13, 18, 21, **22**, 58–60, 153
bin Laden, Osama, 48, 81–82, 217, 218fn12, 219
Blair, Dennis C. (Adm, USN), 63, 64fn11
Boko Haram, 171, 176, 197, 207
  See Nigeria
Brennan, John O., 4, 11, 60fn4, 70–71, 86
  See Central Intelligence Agency
building partner capacity (BPC), 178–79, 181fn17
Bureau of Investigative Journalism (BIJ), 7, 8–**10**, 11, 24–25, **26**–27, 133–**35**, **136**, 137 139
Bush, George W., 87

casualty estimates, 5–19, 83, 195
Center for Army Lessons Learned (CALL), 149
Center for Civilians in Conflict, 149fn19, 158
Central Intelligence Agency (CIA), 1fn1, 48
Chiarelli, Peter W. (LtGen, USA), 22
civilian casualties (CIVCAS), 1–34, 108, **121**, **123**, 138–**39**, 140fn8, 192–96, 204–5

civilian casualty lifecycle, 21–**22**
Civilian Casualty Mitigation Report, U.S. Army, 153
Civilian Casualty Mitigation Team, 204
civilian casualties (CIVCAS), 1–34, 108, **121**, **123**, 138–**39**, 140fn8, 192–96, 204–5
civilians killed (CIV K), 25, 134, 138
civilian protection, xi–xii, 162, 165, 167, 188, 190, 192, 194, 198, 207
Clapper, James R., Jr. (LtGen, USAF), 214–16
Colombia, South America, 180–84, 197, 200, 201fn4, 202, 206
  See Plan Colombia
Commander, International Security Assistance Force (COMISAF), 22–23, 112, 148, 193
consequence management, 28–29, 110, 142, 149fn19, 152
  Conflict Victims Support Project, 28–29
  legal culpability, 28, 152
  restitution, 28–29
Conway, James T. (Gen, USMC), 15
Copenhagen Process, 206–7
  See conduct of hostilities
counterinsurgency (COIN), 112, 189–94
counterterrorism (CT), xi–xii, 2–4, 6fn1, 13, 18fn15, 27, 29fn6, 32fn2, 39, 40–42, 44, 46, 55, 65–68, 71–72, 78, 83, 87, 89, 91–92, 105–68, 171, 178, 186, 189–90, 194–96, 204–5, 209

224 | Index

Creech Air Force Base, Indian Springs, NV, 144

defragmentation, 159–60, 166
Deh Bala District, Nangarhar Province, Afghanistan, 13–14, 17
Department of Defense (DOD), 46–48, 59–63, 69–70, 75–77, 80–82, 96, 166–67, 185, 192, 202, 204, 210–12, 214–17
  See Foreign Military Sales program, Title 10, and Title 50
Department of Justice (DOJ), 52–53, 60, 72–73
directly participate in hostilities (DPH), 6fn2, 68
Director of National Intelligence (DNI), 63, 210, 214
doctrine, 21, 32, 47, 48fn5, 59, 95, 113fn19, 125, 146, 160, 162, 173, 193, 202fn7, 203–4, 214–17
double tapping, 2
drone campaign, U.S., 6fn1, 7–8, 13, 17–18, 24–30, 32–33, 101
  clandestine, 48, 87, 110, 214–16, 220–22
  collateral damage, 2, 4, 20, 38, 52, 58, 60–62, 68fn9, 71, 82–85, 90, 93, 97, 148, 181, 185, 190, 192–93, 195. *See* legitimacy
  covert, 47–49, 56, 61–62, 75, 79, 81, 87, 96fn1, 110, 205, 209–11, 213–22. *See* Title 10, Title 50
  signature strikes, 2, 40, 55, 59, 76–79, 189
  targeting practices, 52, 54, 63, 68. *See* international humanitarian law, international human rights law
drone court, 46, 49–52, 56
  FISA-like drone court, 50, 52, 56, 60, 62, 64, 81, 85, 88, 96–98
  Israeli-style drone court, 51, 56, 60, 62, 64, 80–85, 88, 98–100
drones
  AGM-114 Hellfire air-to-surface missile (ASM), 21fn2, 132, 142
  General Atomics MQ-1 Predator, xi, 16, 141–42, 144–46
  General Atomics MQ-9 Reaper missile, xi
  Patriot missile, 147fn15
  Tomahawk land attack missile (TLAM), 42

enhanced interrogation techniques, 189–90
escalation of force (EOF), 22, 113fn19, 161–62, 192–93
ethical considerations, 2, 84–85, 88, 101

Federal Bureau of Investigation (FBI), 50, 81
Federally Administered Tribal Areas (FATA), 9, 11
  See Pakistan
Feinstein, Dianne, 11, 50, 211

Index | 225

Foreign Intelligence Surveillance Act (FISA), 50
See drone courts
Foreign Military Sales (FMS) program, 202
framework, 37–103, 106, 116, 120, 126–27, 156, 163–64, 220
Fuerzas Armadas Revolucionarias de Colombia (FARC), 181fn15, 182

Gang of Eight, 211
See Feinstein, Diane
Gates, Robert M., 51
Geneva Conventions, xi, 2, 6fn2, 17, 66fn1, 153, 175fn1, 189fn1
See international humanitarian law, international human rights law
Grassley, Charles E., 50
Guantanamo Bay Naval Base, Cuba, 38, 189–90
guilt by association, 16, 145

Hamas, 184–85
Ham, Carter F. (Gen, USA), 180
Heinrich, Martin T., 53
Home Guards Program, 181
See Colombia
host nations, 109
human intelligence (HUMINT), 18, 154
human rights, 71–72, 158, 163, 176, 179–82
See legitimacy
Human Rights Watch, 158

Hurlburt Field, FL, 144

improvised explosive device (IED), 16, 110
insurgent math, 108–9
See McChrystal, Stanley A.
Intelligence Authorization Act, 2014, 53
intelligence collection, 1, 47, 54–56, 62, 91, 210–11
See Title 50
Intelligence Committees, House and Senate, 11, 61, 80, 96, 211–12
See oversight
Intelligence Community (IC), 151, 159, 210
International Armed Conflict (IAC), 67, 74
International Committee of the Red Cross (ICRC), 67, 125, 167
international humanitarian law (IHL), 2, 6, 17fn12, 68, 73–76, 79, 84, 95, 133, 145, 153–54, 160–62, 166–67, 175, 180, 186–91, 194, 198–99, 201–7
See legitimacy
international human rights law (IHRL), 68–69, 72–73, 219
See targeting standards
International Security Assistance Force (ISAF), 7fn3, 15, 22–23, 110, 112, 142, 145, 148, 152, 193–94, 204
imminent, 3, 31, 55–56, 69, 72–73, 76, 78–79, 97, 105, 142, 146, 155, 163, 194

Iraq, 33, 107, 110, 161–62, 165, 171, 175–76, 183–84, 189–94, 197–207
See Islamic State of Iraq and the Levant
irregular enemy, 12–13, 15
Islamic State of Iraq and the Levant (ISIL), 107, 111, 162, 171, 176–77, 194, 197, 201, 207
Israel-Palestine conflict, 184

J7 Joint Force Development, 32–33, 173, 186fn28
Johnson, Jeh C., 69
Joint Civilian Casualty Study (JCCS), 21, 148, 153
Joint Lessons Learned Information System (JLLIS), 185fn27
Joint Staff Civilian Casualties Working Group, 33, 193, 204
joint targeting cycle, 59–60

Karzai 12 rules, 193
Karzai, Hamid, 193
Ki-moon, Ban, 185
King, Angus S., 50
Koh, Harold, 69

law of armed conflict (LOAC), 71–72, 74, 166, 175, 189
Leahy, Patrick J., 50, 201fn4
legal considerations or issues, 2, 41, 71–77, 203
legitimacy, 65–88, 171–207

accountability, 39, 45, 49, 51, 62–63, 77, 82–84, 96, 98–99, 101, 180–81, 202fn7, 203–5, 211. See drone courts, international humanitarian law
domestic, 172, 182
ethical standing, 2, 65, 77, 84–85, 87–88, 90, 101, 166–67, 180, 203
international, 172, 197
supercompliance, 161, 167, 186–87, 198–99, 206
transparency, 7–8, 32, 39–40, 42, 45, 49, 52–53, 63–65, 76–83, 86–87, 90, 95–101, 163, 166, 197, 205–6, 212. See drone courts, oversight
lethal force, xi, 69–70, 87fn58, 105–11, **115**, 117–18, 133, 137, 151, 153, 155–56, 159–60, 162, 164, 167, 204
See counterterrorism operations
Libya, Africa, 23, 107fn6, 161, 165, 171
See Operation Unified Protector

Mali, Africa, 171, 176–80, 197, 207
See Tuareg
McChrystal, Stanley A. (Gen, USA), 109, 157, 193fn8
McKiernan, David D. (Gen, USA), 15, 112
McNeill, Dan K. (Gen, USA), 112
military preference policy, 46–49, 56, 60–63, 75–76, 79–82, 85, 87–88, 95–96, 101, 209–10, 212–13, 219, 221–22
See Title 10, Title 50

Index | 227

military principles
    distinction, 68, 107fn5, 117, 146, 186, 222
    necessity, 56, 68, 73
    proportionality, 68, 84, 186, 188
misidentification, 12–13, 15–19
    See civilian casualties
mission effectiveness or success, 32, 105–13, 114–49, 156, 163–66, 186, 194, 205
Mosul, Iraq, 177
Multinational Force-Iraq (MNF-I), 192fn7, 193

national policy decisions, xi, 156, 186
National Security Agency (NSA), 50
National Security Council, 151, 159
national security practices, xi
nation containing the target (NCT), 92–94
net effectiveness, 40–41, 45, 57, 89–95
New America Foundation (NAF), 7–**10**, 11, 24–**25**, **26**–27, 128, **129**–**31**, 133
Nigeria, Africa, 171, 176, 197, 207
    See Boko Haram
nongovernmental organization (NGO), 5, 29, 125, 149fn19, 163, 167, 192
Noninternational Armed Conflict (NIAC), 67–68, 74
North Atlantic Treaty Organization (NATO), 23, 161

Obama, Barack H., 4, 7, 38, 40–41, 46, 48–49, 51–52, 55, 60, 63, 68–70, 89, 95, 99fn8, 160, 163, 198–99, 217
    See Presidential Policy Guidance
Office of the Secretary of Defense (OSD), 162, 167
operational military effectiveness (OME), 44–45, 57, 62, 89fn3, 94
    See framework, policy options
operational preparation of the environment (OPE), 211–12
Operation Enduring Freedom (OEF), 7fn3, 31–32, 38, 186
Operation Iraqi Freedom (OIF), 147fn15
Operation Unified Protector, 161, 165
    See Libya
operational tempo, 116, 119, 128
operations
    advise-and-assist, 181, 200
    capture, 55–56, 69–70, 75, 85, 91, 107, 155, 189, 191, 220
    clandestine, 48, 87, 110, 214–16, 220–22. *See* Title 10, Title 50
    counterinsurgency, 112, 189–90, 192–94
    counternarcotics, 181. *See* Colombia
    covert, 205, 213–22
    counterterrorism, xi–xii, 2–4, 6fn1, 13, 18fn15, 27, 32fn2, 39–42, 44, 46, 55, 65–66, 68, 71–72, 78, 83, 87, 89, 91–92, 171, 178, 186, 189–90, 194–96, 204–5, 209. *See* Presidential Policy Guidance, Title 10, Title 50

doctrinal covert, 217–18, 220, 222
espionage, 213
lethal force, 106–11, 151
other government agency (OGA), 31, 46–48, 61–62, 74–77, 79–82, 96, 101, 151, 160, 162, 167, 212, **217**, 221
See military preference, Title 50
oversight, 28, 32, 42, 46, 49–52, 56, 58, 61–62, 80–82, 96–98, 151, 155, 160, 165fn2, 166, 205, 209–12, 221
See drone courts, Title 10, Title 50

Pakistan, 1–36, 38, 72, 105, 107, 111–12, 127–40, 153, 165, 171, 189, 194–96, 198, 204–5, 213, 219
See counterterrorism operations
Panetta, Leon E., 81–82, 216–17
pattern of life, 1, 21, 62, 111, 122, 132
Philippines, 180–83, 197, 200–2, 206
Plan Colombia, 181–82, 184
platform precision, 20–23
policy options, 37–103, 176
positive identification (PID), 12, 147, 191
Predator crew, 16, 141–42, 144–46
Presidential Policy Guidance (PPG), 32fn2, 111–12, 114–**15**, 116–18, 122–**23**, 124, 133–34, 136–**40**, 147, 156, 160, 164, 166, 186
prisoner of war (POW), 75
processing, exploitation, and dissemination (PED), 145, 147
See Predator drone

protections, 75, 79, 95. *See* international humanitarian law
combatant privilege, 75
prisoner of war (POW), 75

quick reaction force (QRF), 14

recommendations, 30–35, 95–103, 164–69, 197–207
rules of engagement (ROE), 21, 145–46, 183, 193fn8, 203–4
Russia, 171–72

Saudi Arabia, 171, 188
See counterterrorism operations
security, xi–xii, 33, 41, 52, 54, 74, 86, 91, 93–94, 106, 111, 164, 171–207, 213, 215, 220
Sgrena, Giuliana, 192
See Iraq
Shewan, Farah Province, Afghanistan, 14–15
signature strikes, 2, 40, 55, 59, 76–77, 79
Somalia, Africa, 38, 105, 171
See al-Qaeda, lethal force
Special Operations Forces (SOF), 16, 42, 47, 81, 112, 149, 220
strategic military effectiveness (SME), 44–45, 57, 62, 89fn3, 94

Tactical Directive, COMISAF, 21, 23, 112–13, 148–49, 193

Index | 229

tactical military effectiveness (TME), 40, 44–45, 57–62, 64, 84, 89–98
tactics, techniques, and procedures (TTP), 75, 147
Taliban, 3, 14–16, 110, 142–43, 194
targeting standards, 54, 72–74
target, types, 116–18, 128–**29**, 130–**31**, 132, 145
Title 10, U.S. Code, 47–49, 55–56, 61, 75, 79–81, 88, 96, 159, 209–13, **217**–22
Title 50, U.S. Code, 47–48, 55–56, 61, 75, 81, 96fn1, 159, 209–**17**, 218, 220
See covert operations
Title 50 covert, **217**–18, 220–22
traditional military activity (TMA), 47–48, 55, 75, 96fn1, 210–12, 217–20, 222
Tuareg, 171, 176–78, 207
See Mali

Udall, Mark E., 53
Uniform Code of Military Justice (UCMJ), 203
United Nations (UN), 5, 7–10, 17, 23, 39, 53, 66fn3, 66–67, 71–72, 77, 80, 83, 98, 125, 158, 163, 167, 185, 188, 194
United Nations Security Council Resolution (UNSCR), 66fn3, 67fn4
unmanned aerial system (UAS), 72
unmanned aerial vehicles (UAV), xi, 1, 33, 37–38, 141, 157–58
Uribe, Álvaro, 181–82
See Plan Colombia
Uruzgan Province, Afghanistan, 16, 140–47
U.S. Africa Command (AFRICOM), 180
U.S. Agency for International Development (USAID), 28
U.S. Central Command (CENTCOM), 15
U.S. military group (MILGROUP), 183

Warren, Elisabeth A., 106fn3
See counterterrorism operations
War Powers Resolution, 39, 209fn1
See Authorization for the Use of Military Force
Wyden, Ronald L., 53

Yemen, 3, 11, 18, 27, 29, 32fn2, 38, 54, 61, 105, 107–8, 111–12, 127–40, 153, 158, 165, 171, 188–89, 194–96, 204–5
See civilian casualties, counterterrorism operations